Structural Integrity

Volume 13

Series Editors

José A. F. O. Correia, Faculty of Engineering, University of Porto, Porto, Portugal
Abílio M. P. De Jesus, Faculty of Engineering, University of Porto, Porto, Portugal

Advisory Editors

Majid Reza Ayatollahi, School of Mechanical Engineering, Iran University of Science and Technology, Tehran, Iran
Filippo Berto, Department of Mechanical and Industrial Engineering, Faculty of Engineering, Norwegian University of Science and Technology, Trondheim, Norway
Alfonso Fernández-Canteli, Faculty of Engineering, University of Oviedo, Gijón, Spain
Matthew Hebdon, Virginia State University, Virginia Tech, Blacksburg, VA, USA
Andrei Kotousov, School of Mechanical Engineering, University of Adelaide, Adelaide, SA, Australia
Grzegorz Lesiuk, Faculty of Mechanical Engineering, Wrocław University of Science and Technology, Wrocław, Poland
Yukitaka Murakami, Faculty of Engineering, Kyushu University, Higashiku, Fukuoka, Japan
Hermes Carvalho, Department of Structural Engineering, Federal University of Minas Gerais, Belo Horizonte, Minas Gerais, Brazil
Shun-Peng Zhu, School of Mechatronics Engineering, University of Electronic Science and Technology of China, Chengdu, Sichuan, China

The *Structural Integrity* book series is a high level academic and professional series publishing research on all areas of Structural Integrity. It promotes and expedites the dissemination of new research results and tutorial views in the structural integrity field.

The Series publishes research monographs, professional books, handbooks, edited volumes and textbooks with worldwide distribution to engineers, researchers, educators, professionals and libraries.

Topics of interested include but are not limited to:

- Structural integrity
- Structural durability
- Degradation and conservation of materials and structures
- Dynamic and seismic structural analysis
- Fatigue and fracture of materials and structures
- Risk analysis and safety of materials and structural mechanics
- Fracture Mechanics
- Damage mechanics
- Analytical and numerical simulation of materials and structures
- Computational mechanics
- Structural design methodology
- Experimental methods applied to structural integrity
- Multiaxial fatigue and complex loading effects of materials and structures
- Fatigue corrosion analysis
- Scale effects in the fatigue analysis of materials and structures
- Fatigue structural integrity
- Structural integrity in railway and highway systems
- Sustainable structural design
- Structural loads characterization
- Structural health monitoring
- Adhesives connections integrity
- Rock and soil structural integrity.

Springer and the Series Editors welcome book ideas from authors. Potential authors who wish to submit a book proposal should contact Dr. Mayra Castro, Senior Editor, Springer (Heidelberg), e-mail: mayra.castro@springer.com

More information about this series at http://www.springer.com/series/15775

George Vachtsevanos · K. A. Natarajan ·
Ravi Rajamani · Peter Sandborn
Editors

Corrosion Processes

Sensing, Monitoring, Data Analytics,
Prevention/Protection, Diagnosis/Prognosis
and Maintenance Strategies

Editors
George Vachtsevanos
School of Electrical and Computer Engineering
Georgia Institute of Technology
Atlanta, GA, USA

Ravi Rajamani
drR2 Consulting, LLC
West Hartford, CT, USA

K. A. Natarajan
Department of Materials Engineering
Indian Institute of Science
Bangalore, Karnataka, India

Peter Sandborn
Department of Mechanical Engineering
University of Maryland
College Park, MD, USA

ISSN 2522-560X ISSN 2522-5618 (electronic)
Structural Integrity
ISBN 978-3-030-32830-6 ISBN 978-3-030-32831-3 (eBook)
https://doi.org/10.1007/978-3-030-32831-3

© Springer Nature Switzerland AG 2020
This work is subject to copyright. All rights are reserved by the Publisher, whether the whole or part of the material is concerned, specifically the rights of translation, reprinting, reuse of illustrations, recitation, broadcasting, reproduction on microfilms or in any other physical way, and transmission or information storage and retrieval, electronic adaptation, computer software, or by similar or dissimilar methodology now known or hereafter developed.
The use of general descriptive names, registered names, trademarks, service marks, etc. in this publication does not imply, even in the absence of a specific statement, that such names are exempt from the relevant protective laws and regulations and therefore free for general use.
The publisher, the authors and the editors are safe to assume that the advice and information in this book are believed to be true and accurate at the date of publication. Neither the publisher nor the authors or the editors give a warranty, expressed or implied, with respect to the material contained herein or for any errors or omissions that may have been made. The publisher remains neutral with regard to jurisdictional claims in published maps and institutional affiliations.

This Springer imprint is published by the registered company Springer Nature Switzerland AG
The registered company address is: Gewerbestrasse 11, 6330 Cham, Switzerland

Contents

1. **Introduction** 1
   George Vachtsevanos
2. **Principles of Corrosion Processes** 27
   K. A. Natarajan
3. **Corrosion Sensing** 83
   Jeff Demo and Ravi Rajamani
4. **Corrosion Prevention** 105
   Michael Casey Jones
5. **Data Analytics for Corrosion Assessment** 119
   George Vachtsevanos
6. **Corrosion Modeling** 163
   George Vachtsevanos
7. **Corrosion Diagnostic and Prognostic Technologies** 231
   George Vachtsevanos
8. **Assessing the Value of Corrosion Mitigation in Electronic Systems Using Cost-Based FMEA—Tin Whisker Mitigation** 313
   R. Bakhshi, P. Sandborn and E. Lillie

Contributors

R. Bakhshi University of Maryland, College Park, USA

Jeff Demo Luna Innovations, Roanoke, VA, USA

Michael Casey Jones USAF AFMC AFLCMC/EZP Product Support Engineering Division, AFLCMC/EZPT-CPCO, Corrosion Prevention and Control Office, Robins AFB, GA, USA

E. Lillie University of Maryland, College Park, USA

K. A. Natarajan Department of Materials of Engineering, Indian Institute of Science, Bangalore, India

Ravi Rajamani drR2 Consulting, West Hartford, Connecticut, USA

P. Sandborn University of Maryland, College Park, USA

George Vachtsevanos Georgia Tech, Atlanta, USA

Chapter 1
Introduction

George Vachtsevanos

Abstract This treatise is a comprehensive coverage of corrosion processes addressing the spectrum from the electrochemical fundamentals of corrosion processes to monitoring, sensing, prevention and protection of systems exposed to corrosive processes, data/image mining methods to extract and select useful information from raw data, early and accurate diagnosis of corrosion events, prediction of their time evolution, culminating in maintenance practices of critical assets. Numerous books were published over the last years exploring specific aspects of corrosion processes, monitoring and sensing, prevention and protection materials and processes, condition based maintenance practices, etc. A typical sample of books on corrosion processes published over the recent past is shown below.

G. Vachtsevanos (✉)
Georgia Tech, Atlanta, USA
e-mail: gjv@ece.gatech.edu

© Springer Nature Switzerland AG 2020
G. Vachtsevanos et al. (eds.), *Corrosion Processes*, Structural Integrity 13,
https://doi.org/10.1007/978-3-030-32831-3_1

The challenge is to provide to the expanding corrosion community with a treatise that succinctly and thoroughly covers the most relevant topics with sufficient case studies and references. The book will be useful to the corrosion engineering community, the practitioner, the student, industry and government personnel involved in corrosion assessment and remediation.

Topics covered begin with a thorough introduction to corrosion processes, their impact to sustainment of critical aerospace and industrial processes, typical examples of major corrosion problem areas; the second chapter presents a thorough review of fundamental corrosion processes highlighting the electrochemical nature of these processes; we proceed next to discuss contributions in the corrosion sensor and sensing strategies domain, the Achilles' heel of corrosion assessment seeking major advances in this area; corrosion prevention and protection have been targeted over the past years as those technological advances that with the useful life of important assets; diagnosis and prognosis of the corrosion initiation and evolution are presented next and the book concludes with a treatment of maintenance practices for critical military and industrial systems/processes subjected to corrosion; ROI issues are briefly debated.

1.1 Introduction

Corrosion, in its different forms, is a significant challenge affecting the operational integrity of a vast variety of equipment and processes. Corrosion prevention costs are amounting to billions of dollars each year. As complex equipment age, exposure to corrosion processes is increasing at a substantial and alarming rate contributing to equipment degradation leading to failure modes. Over the past years, it has usually been the high fatality spectacular catastrophic accidents that have worked as the catalyst for change. Historical evidence suggests that fatigue due to corrosion cracking is a major contributor to aircraft accidents. Cracking of critical aircraft structures may endanger severely the performance and life of the vehicle. Corrosion damage can sometimes be greatly exaggerated by the circumstances. While many of the accidents due to failed corroded components have gone non-public for reasons of liability or simply because the evidence disappeared in the catastrophic event, others have made the headlines. The structural failure on April 28, 1988 of a 19-year-old Boeing 737, operated by Aloha airlines, was a defining event in creating awareness of aging aircraft in both the public domain and in the aviation community. Numerous other aircraft catastrophic events were attributed to corrosion accelerated fatigue as the failure mechanism. Figure 1.2 depicts pictures of aircraft catastrophic events attributed to corrosion.

Recent events have demonstrated the importance of early and accurate detection and prediction of the severity and impact of corrosion-induced cracking and the need for immediate remediation/prevention to avoid catastrophic consequences or increased financial burden. In the recent past, cracks on aircraft structures detected during regular maintenance have necessitated urgent actions to be taken to improve the design and installation of replacements for failing components. The pictures in Fig. 1.1 shows the catastrophic effects of corrosion and cracking. Many of these incidents were attributed to corrosion/cracking fatigue. Figure 1.2 is a picture taken

Fig. 1.1 Catastrophic effects of corrosion and cracking (media photos)

Fig. 1.2 Picture of boiler explosion due probably to caustic cracking

decades ago depicting the aftermath of a boiler explosion, probably caused by caustic cracking. Picture courtesy of IMechE.

Corrosion remediation begins with the ability to sense accurately and expeditiously corrosion initiation and growth. Corrosion sensors must be capable of monitoring global and localized corrosion events even in inaccessible regions of aircraft and other systems. Early detection implies corrective actions that will extend the useful life of components/systems exposed to environmental hazardous conditions. A systematic, thorough and robust corrosion modeling effort, addressing all corrosion stages for aluminum alloys or other metals, from micro to meso and macro levels, combined with appropriate sensing, data mining and decision support tools/methods (diagnostic and prognostic algorithms) may lead to substantially improved structural component (materials, coatings, etc.) performance and reduced exposure to detrimental consequences. Reliable, high-fidelity corrosion models form the foundation for accurate and robust corrosion detection and growth prediction. A suitable modeling framework assists in the development, testing and evaluation of detection and prediction algorithms. It may be employed to generate data for data-driven methods to diagnostics/prognostics, test and validate routines for data processing tool development, among others. The flexibility provided by a simulation platform, housing appropriate detection and progression models, is a unique attribute in the study of how corrosion processes are initiated, evolving and may be, eventually, mitigated in physical systems.

Figure 1.3 depicts an integrated framework for corrosion sensing, diagnosis and degradation prediction, impact of corrosion on system integrity and mitigation. We detail the enabling technologies in this book. We highlight corrosion prevention and protection technologies intended to safeguard the integrity of the targeted system by limiting or eliminating surface exposure to corrosion inducing agents (humidity, temperature, etc.). We define a **severity index** resulting from the application of verifiable data mining, diagnostic/prognostic algorithms in real time on-platform aimed to indicate when cracking must be attended to in order to extend the life of critical components, reduce the cost of corrosion prevention and avoid detrimental events. These developments are coupled with current research efforts aiming to design and implement on-platform a "smart" sensing modality that will perform all necessary functions from early detection to prognosis and estimation of the severity of such events. We will rely on a reasoning paradigm built from past historical

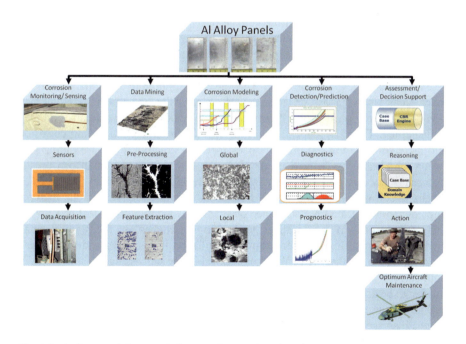

Fig. 1.3 An integrated framework for corrosion sensing, detection/prediction and mitigation

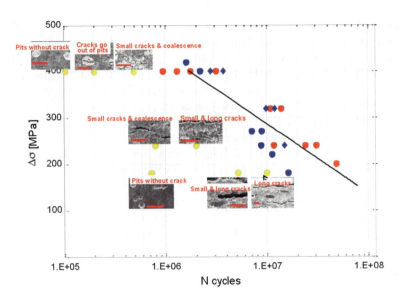

Fig. 1.4 From pitting to cracking of corroding specimens (*source* Dr. Vinod Agarwala)

evidence, learning and adaptation capabilities to assess the severity of cracking and assign an index to the current situation (Fig. 1.4).

Of particular interest to our theme is localized corrosion and cracking, i.e. cracking initiating at points on the surface of a specimen (joints, fasteners, bolts, etc.). A metal surface (aluminum alloy, etc.) exposed to a corrosive environment may, under certain conditions experience attack at a number of isolated sites. If the total area of these sites is much smaller than the surface area then the part is said to be experiencing localized corrosion. The rate of dissolution in this situation is often much greater than that associated with uniform corrosion and structural failure may occur after a very short period. Several different modes of localized corrosion may be identified. These are dependent on the type of specimen undergoing corrosion and its environment at the time of attack. Most destructive forms are pitting corrosion which is characterized by the presence of a number of small pits on the exposed metal surface, crevice attack and cracking. The rapidity with which localized corrosion can lead to the failure of a metal structure and the extreme unpredictability of the time and place of attack, has led to a great deal of study of this phenomenon. In this localized view, imaging studies are focusing on small areas of the global image where corrosion initiation is suspected and may spread more rapidly than other areas. We exploit novel image processing tools/methods, in combination with other means (mass loss calculations) to identify features of interest to be used in the modeling task, since imaging of corroding surfaces offers a viable, robust and accurate means to assess the extent of localized corrosion. We take advantage of first principle, semi-empirical and empirical approaches to modeling of corrosion cracking processes that constitute the cornerstone for

1 Introduction

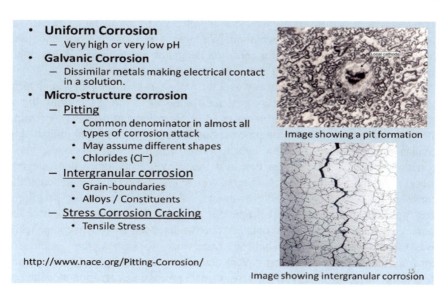

Fig. 1.5 Typical corrosion stages

accurate state assessment and eventual corrective action. It is important to highlight the electrochemical basis for the models, their numerical implementation, and experimental validation and how the corrosion rate of the metal components, at various scales, is influenced by its material properties and the surface protection methods. On the modeling front, a variety of methods has been investigated from data-driven to model-based and empirical or semi-empirical approaches. We present these in detail in the sequel. The evolution of corrosion processes is a crucial step in the assessment process. Figure 1.5 is a pictorial representation of typical corrosion stages.

Micrographs of pitting and cracking corrosion are shown in Fig. 1.6. We study the progression of corrosion through its various stages and employ novel

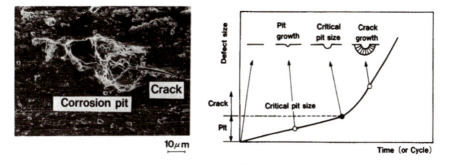

Fig. 1.6 Micrographs of pitting and cracking corrosion; evolution of the corrosion processes [4]

techniques, with performance guarantees, to detect corrosion initiation as early as possible and predict its time evolution. We suggest methods for protection and corrosion prevention. Methods for corrosion mitigation are discussed as they affect important aspects of the overall corrosion assessment strategy.

Major efforts have been underway over the past years to develop and implement corrosion prevention and protection materials/processes to extent the useful life of critical equipment/facilities preventing rapid deterioration and disposal. Assessing the potential impact of corrosion processes on the integrity of critical military and civilian systems, aircraft, ships, industrial and manufacturing processes, transportation systems, etc. requires new and innovative technologies that integrate robust corrosion monitoring, data mining, corrosion detection and prediction of the corrosion (pits, crevices, cracks) growth rate with intelligent reasoning paradigms that capture historical data, expert opinions and adaptation strategies to associate current evidence with past cases obtained fleet-wide for similar system components. We take advantage of a holistic framework to assess the impact of corrosion-induced processes, on typical aluminum alloy components that begins with methods/tools for on-platform sensing, data processing, corrosion modeling of all corrosion stages of particular interest in this study. These functions support diagnostic and prognostic algorithms that are designed to meet customer requirements/specifications for confidence/accuracy and false alarm rates while managing effectively large-grain uncertainty prevalent in health management studies of engineering systems. The hardware/software components of the sensing and health management system form a "smart" sensor that monitors, processes data/ images and decides on-line in real time on the health status and future progression of corrosion pitting/cracking. Corrosion monitoring, detection and prediction entail a series of functions. Starting with the monitoring apparatus, data/image collection and processing, corrosion modeling, detection and prediction and, finally, assessment of the potential impact of corrosion on the operational integrity of an asset. Figure 1.7 depicts the sensing configuration on an aircraft structure. This sequence of corrosion stages is shown schematically in Fig. 1.8. Corrosion states take various forms starting with microstructure corrosion and ending with stress induced cracking.

Corrosion monitoring measures the corrosivity of process conditions by use of appropriate sensors/probes inserted into the process stream exposed continuously to process stream conditions. The nature of the sensors depends on the techniques used for monitoring, accessibility to hidden surfaces, environmental conditions, etc. Corrosion monitoring uses mechanical, electrical, electromechanical devices, among others. They are used on-line in real time or off-line in a laboratory environment. Direct techniques include corrosion coupons, electrical resistance, inductive probes, linear polarization resistance, and impedance spectroscopy, among others, detailed in a separate chapter of this book. The database for corrosion studies consists of coupons, panels, and, sometimes, actual field studies. Figure 1.9 depicts a set of sampled panels, images of cracks and pits and sensing results. There is a need for a considerable database of corrosion data/images to support modeling, diagnostics, prognostics and decision support systems (Figs. 1.10 and 1.11).

1 Introduction

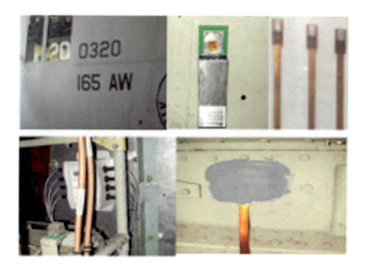

Fig. 1.7 Sensing on aircraft structure

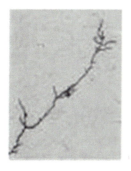  

Fig. 1.8 Corrosion stages

The sensing/modeling and diagnostic/prognostic functions are coupled with a novel reasoning paradigm, called Dynamic Case Based Reasoning (DCBR) that houses a case library composed of past documented cases detailing the impact of cracking on the integrity of platform components/systems. The DCBR is supported by cognitive routines for learning and adaptation so that new evidence is compared with stored cases and those occurring for the first time are "learned" by the reasoner. Figure 1.12 depicts the main modules of the framework. The schematic represents a general architecture for an aircraft corrosion/crack monitoring, the reasoning modules employed to detect and predict the extent of cracking/corrosion,

- Sensing, Temperature, Relative Humidity, Salinity, Mass Loss measurements
- Images of coupons from submersion test and Lap Joint Chamber tests
- Images of cracks and pits found in the literature
- Pictures from field inspection
- Need for on-platform long-term data

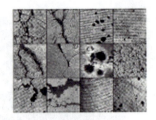

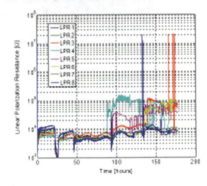

Fig. 1.9 Corrosion data, sample images and polarization resistance measurements, (Isao Shitanda, Ayaka Okumura, Masayaki Itagaki, Kunihiro Watanabe, Yasufumi Asano, Scree-printing atmospheric corrosion monitoring sensor based on electrochemical impedance spectroscopy, https://doi.org/10.1016/snb.2009.03.027)

the life management component and the maintenance actions required. The framework stems from current and past research in corrosion modeling and the development/application of novel CBM+ practices introduced by this research team for military assets. The architecture is set as a decision support system providing advisories to the operator/maintainer as to the health status of critical aircraft component subjected to corrosion and in need of corrective action.

1.1.1 Impact of Corrosion on Engineering System Integrity

It has been established that corrosion is one of the most important factors causing deterioration, loss of metal, and ultimately decrease of nuclear waste management facilities performance and reliability in such critical systems. Corrosion prevention/protection for aging military assets accounts for billions of dollars each year. The situation is similar for commercial enterprises. There is an obvious need to develop and implement new technologies to address these vital issues. Corrosion monitoring, data mining, accurate detection and quantification are recognized as key enabling technologies to reduce the impact of corrosion on the integrity of these

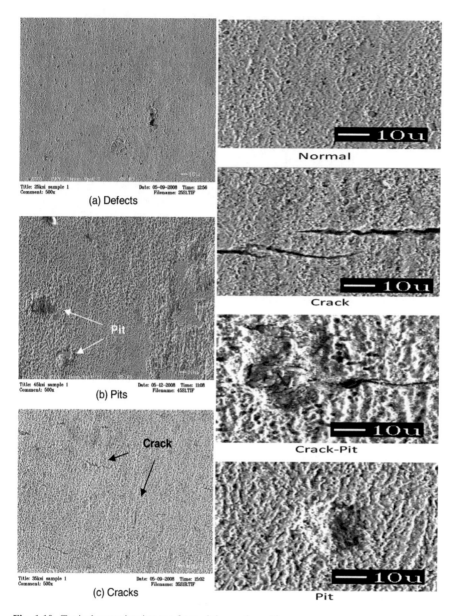

Fig. 1.10 Typical corrosion images from pitting and cracking

assets. Accurate and reliable detection of corrosion initiation and propagation with specified false alarm rates requires novel tools and methods. Corrosion states take various forms starting with microstructure corrosion and ending with stress-induced cracking [1–5].

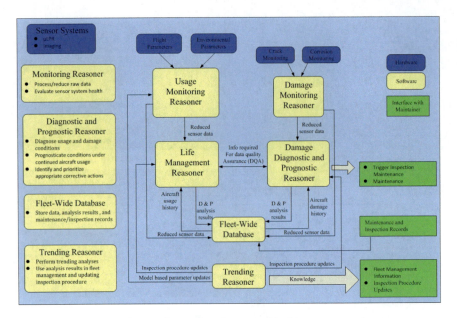

Fig. 1.11 A general architecture for an aircraft structural health management system

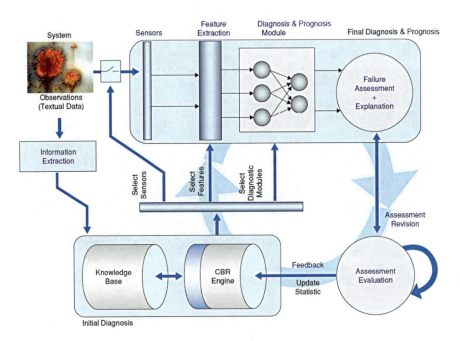

Fig. 1.12 The modules of the dynamic case based reasoning paradigm

1 Introduction

Corrosion cracking (NUREG/CR-7116, SRNL-STI-2011-00005, "Materials Aging Issues and Aging Management for Extended Storage and Transportation of Spent Nuclear Fuel," November 2011. (Available with NRC Accession No. ML11321A182).

Figure 1.13 is a pictorial representation of the corrosion assessment methodology starting with data/images of typical corroded samples and depicting in sequence the sensing, feature or condition indicators extraction and selection, corrosion modeling and knowledge base/classification algorithms.

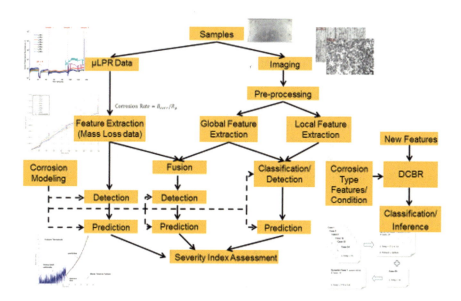

Fig. 1.13 The overall corrosion assessment methodology

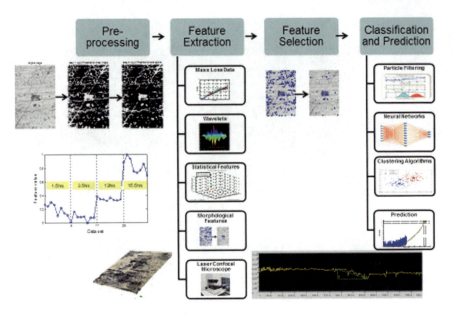

Fig. 1.14 The data/image processing and information extraction architecture

Figure 1.14 depicts the modules of a smart sensing and processing framework. The enabling technologies include means to pre-process raw data/images to improve the signal to noise ratio, feature or characteristic signature extraction and selection to reduce the data dimensionality while preserving the useful information, and classification methods aiming to derive detailed knowledge regarding the type and extent of corrosion.

The framework begins with data/image acquisition and processing and includes all aspects of corrosion detection, prediction, assessment of the impact of corrosion on the integrity of the asset and, finally, corrective action.

Generally speaking, corrosion starts in the form of pitting, owing to some surface chemical or physical heterogeneity, and then facilitated by the interaction of the corrosive environment fatigue cracks initiate from corrosion pitted areas and further grow into the scale that would lead to accelerated structure failure [6, 7]. In order to effectively conduct structural corrosion health assessment, it is thus crucial to understand how corrosion initiates from the microstructure to the component level and how structure corrosion behaviors change as a result of varied environmental stress factors. Many research efforts have been reported in the past addressing this critical issue [8–11]. Traditionally, conventional ultrasonics and eddy current techniques have been used to precisely measure the thickness reduction in aircraft and other structures; there has been a number of undergoing research using guided wave tomography techniques to screen large areas of complex structures for corrosion detection, localization [12] and defect depth mapping [13]. However, due to the nature of ultrasonic guided wave, this technique is vulnerable to environmental changes, especially to temperature variation and surface wetness occurrence [14], and the precision of corrosion defect depth reconstruction is restricted by sensor network layout, structure complexity, among others, which limits the scope of the field application. Thus, undeniably, well-recognized global corrosion measurements, such as material weight loss and wall thickness reduction, cannot offer an appropriate and trustworthy way to interpret the pitting corrosion due to its localized attack nature.

Besides, advanced corrosion health assessment systems require comprehensive quantitative information, which can be categorized into a variety of feature groups, such as corrosion morphology, texture, location, among others. It calls for the exploration of both new testing methods and data fusion methods from multiple testing techniques. Forsyth and Komorowski [15] discussed how data fusion could combine the information from multiple NDE techniques into an integrated form for structural modeling. Several other studies have looked into different sensing technologies for corrosion health monitoring, including using a micro-linear polarization resistance, μLPR sensor [16, 17], and fiber optic sensors [18]. However, the existing research effort in a combination of surface metrology and image processing is very limited. In parallel to the current corrosion sensing technology, there have been a number of corrosion modeling studies trying to numerically capture the processes of pitting corrosion initiation, pitting evolvement, pitting to cracking transition, and crack growth to fracture at the molecular level. However, currently there is no accepted quantitative model to take into consideration the effect of stress factors (e.g., salinity, temperature, pressure), although the effects of the above-mentioned stress factors have been widely discussed. Corrosion protection has been attracting research over the past decades. Multidisciplinary efforts in materials, processes, equipment and applications over the past decades have resulted in significant advances. These efforts continue as aging aircraft are in need of repair and maintenance to sustain their operational integrity. Figure 1.15 shows methods, measures and procedures for corrosion protection (von Octeren, Korrosionschutz-Fibel).

(W. von Baeckmann, W. Schwenk and W. Prinz, editors, Corrosion Protection, Theory and Practice of Electrochemical Protection Processes, Third Edition, Gulf Professional Publishing, 1997).

Surface Coating There are two main types of surface coating for corrosion protection: Metallic coating where the structure is coated with a layer of another metal which may be more noble than the structure or less noble than it, for example, steel structures can be coated with copper which is more noble than steel or zinc which is less noble. Certain factors must be considered in selection of a coating metal, such as resistance to direct attack of the environment of the coating metal, must be non-porous and hard, etc. Figure 1.16 shows the main coating materials.

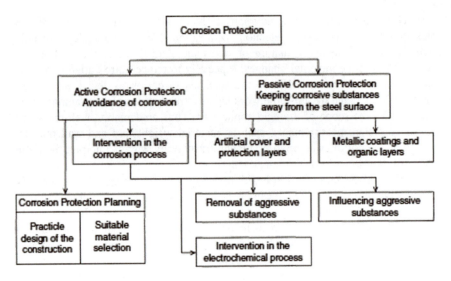

Fig. 1.15 Methods, measures and procedures of corrosion protection (von Octoren, Korrosionsschutz-Fibel)

Fig. 1.16 Coating materials

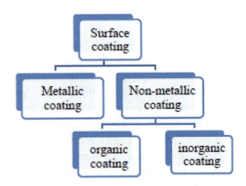

Directives, laws and commands have been issued requiring measures to be taken to reduce the impact of corrosion on critical systems/processes. As an example, the following is directed to the military:

***Public Law 107-314 s: 1067. Prevention and mitigation of corrosion of military infrastructure and equipment* requires that**:

DoD designate a responsible official or organization.
DoD developed a long-term corrosion strategy to include.

- Expansion of emphasis on corrosion prevention and mitigation.
- Uniform application of requirements and criteria for the testing and certification of new corrosion prevention technologies within common materiel, infrastructure, or operational groupings.
- Implementation of programs to collect and share information on corrosion within the DoD.
- Establishment of a coordinated R&D program with transition plans.

Strategy to include policy guidance and assessment of funding and personnel resources required

- Corrosion Policy in
    - DoD 5000 Guidebook—part of systems engineering.
    - Guidebook for designing and assessing supportability in DOD weapon systems.
    - CPC Requirements included in capabilities docs (ICD/CDD).
    - DFARS—corrosion planning required for all programs requiring acquisition plans.
    - DoDI 5000.2.
- CPC part of performance based acquisition and logistics.
- CPC planning guidebook published—Spiral 2.

The Military point of view:
Margery Hoffman and Paul Hoffman, "Current and Future Life Prediction Methods for Aircraft Structures", Naval Research Reviews, vol. 50, No. 4, pp. 4–13, 1998.

- Fatigue Life Expended (FLE):
    - An index relative to test flight hours it takes to form 0.01 cracks. FLEs are calculated at 5–9 locations for fighter/attack aircraft and at 20–30 locations for patrol and support aircraft.
    - FLEs are used to schedule routine maintenance and structural inspections, life assessment prognosis, service life extension programs, and retirements of aircraft from the active fleet.
- Two Major AF Activities:
    1. Analytically determine the service life of an aircraft structure and then validate that life through a full-scale fatigue test.
    2. Setup an individual fatigue-tracking program that collects aircraft usage data and performs fatigue predictions quarterly for every fatigue critical component.

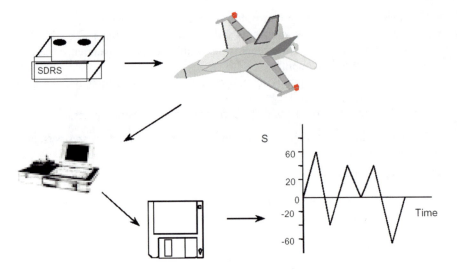

Stress History at critical locations

- The 4 Elements of the FLE Tracking Process:
  - Data Collection.
  - Data Reduction.
  - Damage Calculation.
  - Information Dissemination.

A symbolic representation of the "Total Life" concept, as applied to a fleet of aircraft is shown in Fig. 1.17.

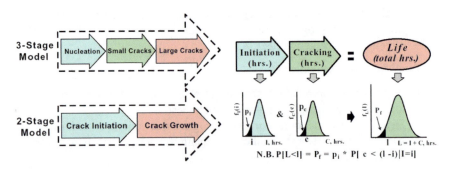

Fig. 1.17 Two different models of "total life"

1.2 Fatigue Corrosion: Example Cases in Aerospace and Industrial Processes

Major industrial processes for corrosion detection include pipeline diagnostics hardware and software methods/tools, as depicted in Fig. 1.18.

Nuclear waste stainless steel disposal facilities have been targeted for corrosion assessment and remediation since they are typically exposed to corrosive environments. EPRI, DoE and other government and industry organizations have been actively seeking methods to monitor waste disposal canisters for corrosion, data collection and analysis as well means to extend the canister's useful life beyond 100 years. Figure 1.19 shows plausible canister degradation mechanisms and the associated key parameters affecting these corrosion mechanisms.

Figure 1.20 is an illustration of ultrasonic sensing to detect standing water in bottom of spent fuel storage containers.

Corrosion and corrosion related factors undermine the fatigue properties of materials used in aircraft construction to a significant extent that service failures are a serious concern. Corrosion in aging aircraft can aggravate metal fatigue to the point where service life is reduced. The Department of Defense spends billions of dollars each year for corrosion repairs and maintenance of 15,000 aircraft. Avoiding corrosion fatigue is a formidable task especially in naval aviation because they operate in the most severe environments. Research studies have been underway for many years to understand and model the metallurgical, mechanical and electrochemical aspects of fatigue corrosion. Parallel efforts focus on the development of appropriate corrosion prevention/protection materials and strategies. (Vinod S.

- Issues
    - Detect defects
    - Polar environment
    - Overlapping defects
    - Continuous data stream

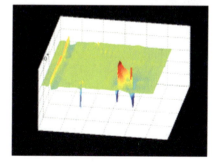

Fig. 1.18 Pipeline inspection processes

Plausible Canister Degradation Mechanism	Key Parameters
CISCC	Deposited chlorides (quantity and associated cation) Presence of water (surface humidity above DRH, rain ingress, etc.) Residual or applied stress Surface temperature Material condition (microstructure, sensitization, and fabrication defects) Composition of surface deposits (e.g., presence of free iron, dust, etc.) Cold work and surface condition (grinding, polishing, etc.) Presence of crevices (macrocrevices and microcrevices due to grinding, etc.)
Pitting Corrosion	Quantity and type of aggressive species (e.g., chlorides) Presence of water (deliquescence above DRH, rain ingress, etc.) Composition of surface deposits (e.g., presence of free iron, dust, etc.) Surface temperature Surface solution pH Material condition (presence of inclusions, sensitization, fabrication defects)
Crevice Corrosion	Occluded area (geometry or impermeable deposit) Presence of water (surface humidity above DRH, rain ingress, etc.) Quantity and type of aggressive species (e.g., chlorides, graphite) Surface temperature Crevice solution pH
Microbiologically Induced Corrosion	Presence of water or very high relative humidity Source of nutrients (CO_2, dust, etc.) Radiation resistant microbes Deposition of bacterial colony Low surface temperature
Intergranular Attack	Presence of water (surface humidity above DRH, rain ingress, etc.) Very low pH solution Sensitized microstructure

Source: Failure Modes and Effects Analysis (FMEA) of Welded Stainless Steel Canisters for Dry Cask Storage Systems, 3002000815, EPRI, Final Report, December 2013

Fig. 1.19 Nuclear waste canisters: degradation mechanisms and key parameters causing degradation

Fig. 1.20 Illustration of ultrasonic sensing to detect standing water in bottom of spent fuel storage container

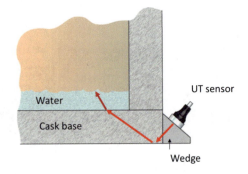

Agarwala, Fatigue and Corrosion: Aircraft Concerns, Naval Research Reviews, Volume 50, November 4, 1998).

The fundamental constraint for Navy aircraft is that the broad theater of operations of aircraft carriers and the limited space on board for maintenance restrict routine inspections for fatigue cracks. (Margery E. Hoffman and paul C. Hoffman, Current and Future Fatigue Life Prediction Methods for Aircraft Structures, Naval Research Reviews, Volume 50, November 4, 1998).

1 Introduction

1.3 Corrosion of Oil Platforms

Corrosion in steel oil platforms can lead to damage and failure of the structure resulting in expensive repairs, loss of business and even on-site accidents. Figure 1.21 is a picture of an oil rig while Fig. 1.22 depicts the corrosion process in steel structures under seawater and Fig. 1.23 shows a protective mechanism for undersea structures using a sacrificial anode method.

Fig. 1.21 An oil rig

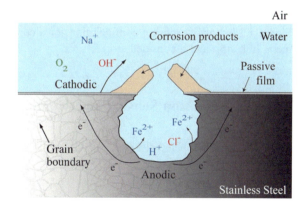

Fig. 1.22 An outline of the corrosion process for a steel structure in seawater. *Image credit* Naval Research Laboratory (NRL)

Fig. 1.23 Sacrificial anodes (the white handle-shaped objects protect the oil platform from corrosion. Image by Chetan, via Wikimedia Commons)

1.4 Pipeline Fatigue Corrosion

Tracking of corrosion fatigue in gas pipelines is a major challenge. Many approved technologies are available for measuring corrosion, corrosion coupons, electrical resistance probes, with most of these technologies measuring the corrosivity of the gas rather than that the changes in the pipeline wall. Fiber optic devices, with associated networked monitoring, overcome many shortcomings of conventional sensors.

1.5 Concrete Block Corrosion Sensing

Figure 1.24 shows a sensing configuration for a concrete block using a sacrificial metal link strip and a corroded strip.

1.6 GE Corrosion Sensing and Monitoring Technologies

Figure 1.25 depicts apparatus and sensing results provided by GE.

1 Introduction

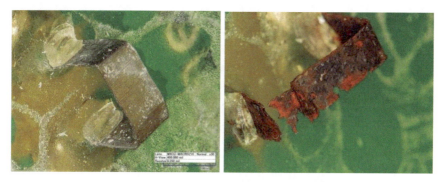

Fig. 1.24 A SWA corrosion sensor removed from a concrete block had an un-corroded sacrificial metal link sensor strip (left) and a corroded strip (right) (photos courtesy of FDOT)

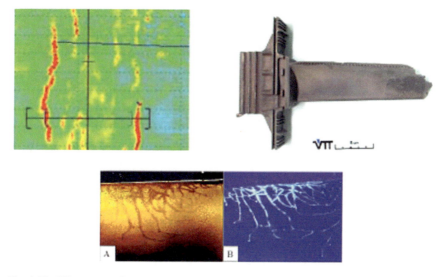

Fig. 1.25 GE apparatus for corrosion sensing and imaging results

1.7 Corrosion of Steel in Concrete Structures

Corrosion of steel in concrete structures is a reason for infrastructure failures. There is a need for effective and robust sensing technologies to detect accurately and expeditiously corrosion initiation and growth so that remedial action can be taken to extent the integrity and life of infrastructure. Assessing the corrosion condition in buried steel is a challenge. Fiber Bragg Grating (FGB) sensors and other sensing modalities have been suggested to address this problem.

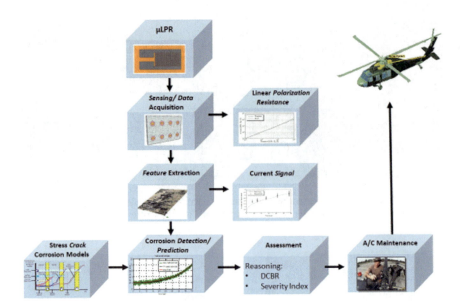

Fig. 1.26 From the laboratory to on-platform realization of corrosion assessment technologies

1.8 Corrosion Assessment: From the Laboratory to On-Board the Aircraft

Figure 1.26 is a depiction of those technologies that realize a monitoring, image interpretation and detection/prediction that must be transitioned from the laboratory environment to on-board the aircraft.

References

1. Wallace W, Hoeppner DW (1985) AGARD corrosion handbook volume I aircraft corrosion: causes and case histories. AGARD-AG-278, vol 1
2. Wei RP, Liao CM, Gao M (1998) A transmission electron microscopy study of 7075-T6 and 2024-T3 aluminum alloys. Metall Mater Trans A 29A:1153–1163
3. Hoeppner DW, Chandrasekaran V, Taylor AMH (1999) Review of pitting corrosion fatigue models. In: International Committee on aeronautical fatigue, Bellevue, WA, USA
4. Kawai S, Kasai K (1985) Considerations of allowable stress of corrosion fatigue (focused on the influence of pitting). Fatigue Fracture Eng Mater Struct 8(2):115–127
5. Lindley TC, Mcintyre P, Trant PJ (1982) Fatigue-crack initiation at corrosion pits. Metals Technol 9(1):135–142
6. Pidaparti RM (2007) Structural corrosion health assessment using computational intelligence methods. Struct Health Monitor 6(3):245–259
7. Rao KS, Rao KP (2004) Pitting corrosion of heat-treatable aluminum alloys and welds: a review. Trans Indian Inst Met 57(6):593–610

8. Frankel GS (1998) Pitting corrosion of metals: a review of the critical factors. J Electrochem Soc 145(6):2186–2198
9. Huang T-S, Frankel GS (2006) Influence of grain structure on anisotropic localized corrosion kinetics of AA7xxx-T6 alloys. Corros Eng Sci Technol 41(3):192–199
10. Szklarska-Smialowska Z (1999) Pitting corrosion of aluminum. Corros Sci 41(9):1743–1767
11. Pereira MC, Silva JW, Acciari HA, Codaro EN, Hein LR (2012) Morphology characterization and kinetics evaluation of pitting corrosion of commercially pure aluminum by digital image analysis. Mater Sci Appl 3(5):287–293
12. Clark T (2009) Guided wave health monitoring of complex structures. Imperial College London, London
13. Belanger P, Cawley P, Simonetti F (2010) Guided wave diffraction tomography within the born approximation. IEEE Trans UFFC 57:1405–1418
14. Li H, Michaels JE, Lee SJ, Michaels TE, Thompson DO, Chimenti DE (2012) Quantification of surface wetting in plate-like structures via guided waves. AIP Conf Proc Am Inst Phys 1430(1):217
15. Forsyth DS, Komorwoski JP (2000) The role of data fusion in NDE for aging aircraft. In: SPIE aging aircraft, airports and aerospace hardware IV, vol 3994, p 6
16. Brown D, Darr D, Morse J, Laskowski B (2010) Real-time corrosion monitoring of aircraft structures with prognostic applications. In: Annual conference of the prognostics and health management society, vol 3
17. Brown DW, Connolly RJ, Laskowski, B, Garvan M, Li H, Agarwala VS, Vachtsevanos G (2014) A novel linear polarization resistance corrosion sensing methodology for aircraft structure. In: Annual conference of the Prognostics and Health Management Society, vol 5, no 33
18. Li H, Garvan M, Li J, Echauz J, Brown D, Vachtsevanos G, Zahiri F (2017) An integrated architecture for corrosion monitoring and testing, data mining, modeling, and diagnostics/prognostics. Intl J Progn Health Manage 8(5):12

Chapter 2
Principles of Corrosion Processes

K. A. Natarajan

Abstract In this chapter, principles of corrosion processes are illustrated with special emphasis on electrochemical aspects. Galvanic and electrolytic cells implicated in various corrosion and protection processes are analyzed with examples. Concentration cells are outlined with reference to galvanic corrosion and formation of differential aeration (oxygen) regions leading to pitting corrosion. Electrochemical kinetics and mixed potential theory are discussed. Active-passive transition in metals and alloys is brought out. Anodic and cathodic protection based on electrochemical principles find technological applications in industrial corrosion protection. Mechanisms involved in biofouling and microbially influenced corrosion are critically analyzed. Human body as a corrosion environment with reference to implanted biomaterials is also brought out.

Corrosion is the deterioration or destruction of metals (and alloys) in the presence of an environment brought about by chemical or electrochemical means.

Microbiologically-influenced corrosion (MIC) has now assumed great significance and can be defined as deterioration or destruction of metals and alloys by electrochemical or biological means in the presence of microorganisms [1–5].

Corrosion types can be classified into dry and wet in general, while the environments can be liquid (aqueous), underground, atmospheric and high temperature. Electrochemical principles come into play in all cases.

Different forms of corrosion can occur depending on wide range of possible environments.

Common industrial classifications are as follows.

(a) Uniform corrosion
(b) Galvanic corrosion
(c) Localized corrosion such as Crevice corrosion, Pitting, Filiform corrosion
(d) Selective leaching as in alloys such as brass (Dezincification)
(e) Intergranular/Transgranular attack

K. A. Natarajan (✉)
Department of Materials of Engineering, Indian Institute of Science, Bangalore 560012, India
e-mail: kan@iisc.ac.in

(f) Erosion, Impingement and Cavitation corrosion
(g) Stress cracking and Stress corrosion cracking (Hydrogen embrittlement, Sulfide stress cracking, Liquid metal embrittlement)
(h) Fretting corrosion, Corrosion fatigue
(i) High temperature oxidation.

Although biological or microbial corrosion cannot be classified under types of corrosion, it has become an industrially relevant, widespread, catastrophic form of corrosion.

Electrochemical and microbial corrosion reactions involve electrochemical cells (corrosion cells) consisting of anode, cathode and an electrolyte and anodic (oxidation) and cathodic (reduction) reactions. When the electrodes are interconnected, a potential difference is developed (see Fig. 2.1).

Anode: Electrode where oxidation (corrosion) occurs

$$M = M^{++} + 2e \qquad (2.1)$$

Cathode: Electrode where reduction (deposition) occurs

$$M^{++} + 2e = M \qquad (2.2)$$

For every oxidation reaction, there is a reduction reaction as well, and the net reaction represents the total electrochemical process.

As an example, in the corrosion of zinc metal in an acid solution, the respective reactions are

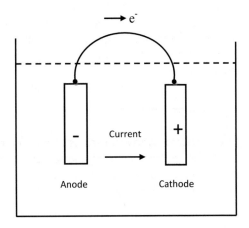

Fig. 2.1 Basic galvanic corrosion cell

$$Zn = Zn^{++} + 2e \quad \text{(Anode)} \tag{2.3}$$

$$\underline{2H^+ + 2e = H_2 \quad \text{(Cathode)}} \tag{2.4}$$

$$\underline{Zn + 2H^+ = Zn^{++} + H_2 \text{ (Net)}} \tag{2.5}$$

Several types of electrodes such as

Metal-metal ion	Fe/Fe^{++}
Ion/Ion (redox)	Pt/Fe^{+++}, Fe^{++}
Gas	Pt/H$_2$, H$^+$
Metal-insoluble salt come into play	Hg/Hg$_2$Cl$_2$, KCl

Electrochemical cells can be divided into galvanic and electrolytic cells and differentiated as follows [1–5].

Galvanic	Electrolytic
• Chemical to electrical energy	• Electrical to chemical energy
• Spontaneous/reversible reactions	• Non-spontaneous/Kinetic cell
• Cathode (+), Anode(−)	• Cathode (−), Anode (+)
• $\Delta G^0 < 0$, $E^0_{Cell} > 0$	• $\Delta G^0 > 0$, $E^0_{Cell} < 0$
• Eg: Dry cell, Daniel Cell	• Eg: Electroplating, Cathodic protection by impressed current

Sign conventions are followed to denote half-cell electrode reactions.

European convention, American convention and International (IUPAC) conventions are followed since long.

In this chapter, IUPAC convention is followed, where the half-reaction is expressed as a reduction reaction and 'Plus Right Rule' used to estimate total cell EMF.

Free energy concepts as thermodynamic fundamentals are used to estimate relationships between

- Free energy change and equilibrium constant
- Free energy change and cell EMF

And the Nernst expression derived as follows

$$E = E^0 + \frac{RT}{nF} \ln \frac{[Ox]}{[Red]} \tag{2.6}$$

$$E = E^0 + \frac{0.059}{n} \log \frac{[Ox]}{[Red]} \text{ (Simplified to room temperature conditions)}$$

where

E = Half-cell potential
E^0 = Standard Electrode Potential

$$\frac{[Ox]}{[Red]} = \text{ratio of activities of oxidised/reduced species.}$$

Daniel cell can be taken as a typical example to illustrate cell EMF calculation and to establish criterion for spontaneity.

Daniel cell consists of zinc and copper electrodes in a diaphragm cell configuration, immersed in 1M $ZnSO_4$ and 1M $CuSO_4$ respectively.

$$-Zn|Zn^{++}_{(1M)}||Cu^{++}_{(1M)}|Cu+$$

$$Zn = Zn^{++} + 2e \quad E^0 = -0.76V \tag{2.3}$$

$$Cu^{++} + 2e = Cu \quad E^0 = +0.34V \tag{2.7}$$

$$Zn + Cu^{++} = Zn^{++} + Cu \tag{2.8}$$

$$E_{Cell} = E_{\frac{1}{2}}(Right) - E_{\frac{1}{2}}(Left)$$

$$E_{\frac{1}{2}}(Zn) = -0.76 + \frac{0.059}{2}\log[Zn^{++}] \tag{2.9}$$

$$E_{\frac{1}{2}}(Cu) = +0.34 + \frac{0.059}{2}\log[Cu^{++}] \tag{2.10}$$

$$E_{Cell} = +0.34 - (-0.76) = +1.10\,V$$

(The above cell is spontaneous).

2 Principles of Corrosion Processes

The driving force for the corrosion reaction is the potential difference (cell EMF) between anode and cathode. In a corroding metal, several anodic and cathodic sites exist.

Since absolute potential of a single electrode cannot be measured, all potential measurements in electrochemical corrosion cells are made relative to a reference electrode.

Basic electrochemical aspects of commonly used reference electrodes are illustrated below.

2.1 Silver–Silver Chloride Reference Electrode

The redox reaction at the electrode is

$$AgCl + e = Ag + Cl^- \qquad (2.11)$$

The electrode consists of a silver wire coated with silver chloride immersed in chloride solution.

$$E = E^0 + \frac{RT}{nF} \log \frac{1}{Cl^-} \qquad (2.12)$$

Variations in chloride ion concentrations influence the potential.

2.2 Saturated Calomel Electrode (SCE)

The redox reaction for the electrode is

$$Hg_2Cl_2 + 2e = 2Hg + 2Cl^- \qquad (2.13)$$

$$E = E^0 + \frac{RT}{nF} \log \frac{1}{[Cl^-]^2} \qquad (2.14)$$

$$E^0 + \frac{0.059}{2} \log \frac{1}{[Cl^-]^2} \qquad (2.15)$$

2.3 The Hydrogen Electrode (NHE)

A platinum wire contacted with an acid solution of unit H^+ activity. Pure H_2 gas at one atmosphere is bubbled into solution at room temperature.

$$2H^+ + 2e = H_2 \tag{2.4}$$

$$E = E^0 + \frac{0.059}{2}\log[H^+]^2 \tag{2.16}$$

$$E = E^0 - 0.059\,pH \tag{2.17}$$

The Standard Hydrogen Electrode (SHE) is also referred to as Normal Hydrogen Electrode (NHE) with the standard potential, $E^0 = 0.00$ V.

2.4 Copper–Copper Sulfate Electrode

Very robust and stable reference electrode, often used in cathodic protection systems to measure pipe to soil potentials. Copper metal is placed in a solution of copper sulfate (saturated)

$$Cu^{++} + 2e = Cu \tag{2.7}$$

$$E_{Cu^{++}/Cu} = 0.34 + \frac{0.059}{2}\log[Cu^{++}] \tag{2.10}$$

The standard potential for the electrode is +0.316 V

Standard potentials of different reference electrodes are summarized below:

System	Electrolyte	E^0 V
Calomel	Sat'd KCl	0.241
$2Hg + 2Cl^- = Hg_2Cl_2 + 2e$	1.0 N KCl	0.280
	0.1 N KCl	0.333
Silver–Silver chloride	Sat'd KCl	0.199
$Ag + Cl^- = AgCl + e$	1.0 N KCl	0.234
	0.1 N KCl	0.288
Copper–Copper sulfate	Sat'd $CuSO_4$	0.316
$Cu^{++} + 2e = Cu$		

Diagrammatic representations of various reference electrodes are given in Figs. 2.2, 2.3, 2.4 and 2.5.

Fig. 2.2 Silver-Silver chloride electrode

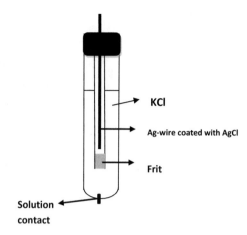

Fig. 2.3 Saturated calomel electrode (SCE)

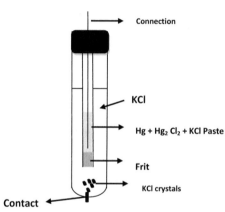

Fig. 2.4 Standard (normal) hydrogen electrode

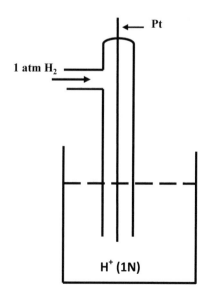

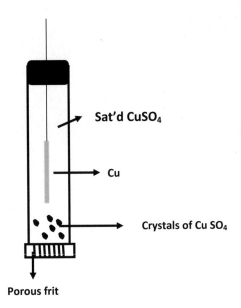

Fig. 2.5 Copper-copper sulfate electrode

2.5 Junction Potentials

A small potential difference that develops at the interface (junction) of two electrodes arises due to different ionic compositions. Liquid junction corrections need to be made to avoid interferences in measured electrode potentials.

2.6 Concentration Cells [1–5]

Besides dissimilar electrode cells (bimetallic), concentration cells can also be considered galvanic corrosion cells. In concentration cells, the EMF arises due to changes in concentrations of the electrolytes as well as reactants such as oxygen. There is no net chemical reaction and the electrical energy arises from the transfer of a reactant in varying concentrations from electrolytes.

For example:

- Differences in metal ion concentrations.
- Differences in oxygen partial pressures.

Example:

$$- \text{Ag} \left| \text{Ag}_{(C_1)} \text{NO}_3 \right| \left| \text{Ag}_{(C_2)} \text{NO}_3 \right| \text{Ag} +$$

2 Principles of Corrosion Processes

c_1 and c_2 are the Ag^+ concentrations in the anode and cathode compartments.

$$E_{Cell} = E_{\frac{1}{2}(R)} - E_{\frac{1}{2}(L)}$$
$$= \left[E^0 + \frac{0.059}{n}\log C_2\right] - \left[E^0 + \frac{0.059}{n}\log C_1\right] \quad (2.18)$$

$$= \frac{0.059}{n}\log \frac{C_2}{C_1} \quad (2.19)$$

The EMF is developed due to transfer of metal ions from the area of higher concentration to that of lower concentration. The silver electrode in contact with a lower silver ion concentration serves as anode, the other being the cathode.

Differential aeration corrosion occurs when a metal surface is exposed to differential air or oxygen concentrations. The part of the metal exposed to higher O_2 concentration acts as cathode, while the part exposed to lower oxygen concentration serve as anodic regions. Poorly oxygenated regions thus undergo corrosion and oxygen-enriched areas are protected,

$$\text{Anode} = M = M^{++} + 2e(\text{Low } O_2) \quad (2.1)$$

$$\text{Cathode} = 1/2\, O_2 + H_2O + 2e = 2OH^- (\text{High } O_2) \quad (2.20)$$

Diagrammatic illustrations of the above concentration cells are given in Figs. 2.6 and 2.7.

Water line corrosion is a case of differential aeration corrosion which is prevalent in sea going vessels, water storage tanks and submerged structures. Oxygen concentration cells are formed in this type of corrosion. The part of the metal below the water line is exposed to lower oxygen levels while the part above the water is

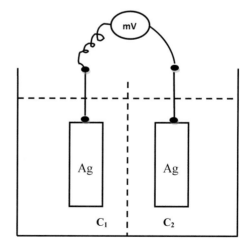

Fig. 2.6 Metal–ion concentration cell

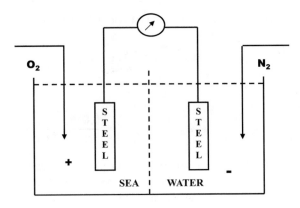

Fig. 2.7 Differential oxygen (aeration) cell

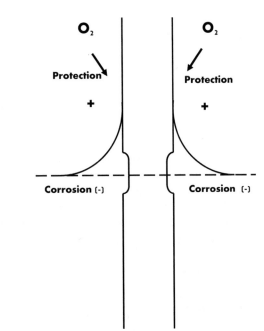

Fig. 2.8 Water-line corrosion (oxygen concentration cell)

exposed to higher oxygen partial pressures and consequently, the metal part below water line acts as anode and undergo corrosion (see Fig. 2.8).

The Nernst relationship can be used to estimate the potential difference generated due to oxygen concentration cells. When two portions of the same metal are in contact with a solution having differential oxygen concentration areas,

$$O_2 + 2H_2O + 4e = 4OH^- \qquad (2.21)$$

2 Principles of Corrosion Processes

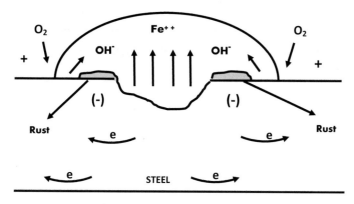

Fig. 2.9 Rusting of iron through oxygen concentration cell formation

$$E_1 = E^0 + \frac{0.059}{4} \log \frac{PO_{2(a)}}{[OH^-]^4} \quad (2.22)$$

$$E_2 = E^0 + \frac{0.059}{4} \log \frac{PO_{2(b)}}{[OH^-]^4} \quad (2.23)$$

$PO_{2(a)}$ and $PO_{2(b)}$ are different oxygen partial pressures while pH being the same for both half cells

$$E_2 - E_1 = \frac{0.059}{4} \log \frac{PO_{2(b)}}{PO_{2(a)}} \quad (2.24)$$

If $PO_{2(a)} < PO_{2(b)}$, then $E_2 > E_1$

Electrode in contact with lower O_2 concentration is anode (corrosion) and the one in contact with higher O_2 is cathode (protection)

A model for role of oxygen concentration cells (differential aeration) on rusting of iron exposed to oxygen and moisture is illustrated in Fig. 2.9.

2.7 EMF Series [1–5]

The EMF series is an arrangement of various metals in order of their electrochemical activities based on their standard electrode potentials. It is a thermodynamic series where the E^0 is calculated from free-energy data. The most active metal in the series possesses the highest negative E^0 (anode) while the nobler metal having less negative or more positive electrode potential (cathode)

There are exceptions to the thermodynamically predicted metal activities as listed in Table 2.1.

Table 2.1 EMF series

Reaction	E^0, V(SHE)	
$Au^{+++} + 3e = Au$	+1.42	Noble
$Pt^{++} + 2e = Pt$	+1.2	↑
$O_2 + 4H^+ + 4e = 2H_2O$	+1.23	
$Pd^{++} + 2e = Pd$	+0.83	
$Ag^+ + e = Ag$	+0.799	
$O_2 + 2H_2O + 4e = 4OH^-$	+0.401	
$Cu^{++} + 2e = Cu$	+0.34	
$Sn^{+++} + 2e = Sn^{++}$	+0.154	
$2H^+ + 2e = H_2$	0.00	Reference
$Pb^{++} + 2e = Pb$	−0.126	
$Sn^{++} + 2e = Sn$	−0.140	
$Ni^{++} + 2e = Ni$	−0.23	
$Co^{++} + 2e = Co$	−0.27	
$Cd^{++} + 2e = Cd$	−0.402	
$Fe^{++} + 2e = Fe$	−0.44	
$Cr^{+++} + 3e = Cr$	−0.71	
$Zn^{++} + 2e = Zn$	−0.763	
$Al^{+++} + 3e = Al$	−1.66	
$Mg^{++} + 2e = Mg$	−2.38	
$Na^+ + e = Na$	−2.71	↓
$K^+ + e = K$	−2.92	Active

Example: Aluminum, chromium and titanium though active in the EMF series are practically highly corrosion resistant due to the presence of stable metal oxide passive surface films.

2.8 Applications of EMF Series

(a) A more active metal would displace a nobler metal from its dissolved state in an aqueous solution. For example, both iron and zinc (being placed lower to copper in the EMF series) can displace cupric ions (reduce) from an acid solution.

$$Zn + CuSO_4 = ZnSO_4 + Cu \qquad (2.25)$$

$$Fe + CuSO_4 = FeSO_4 + Cu \qquad (2.26)$$

$$\text{Oxidation} \quad Fe = Fe^{++} + 2e \qquad (2.27)$$

$$Zn = Zn^{++} + 2e \qquad (2.3)$$

$$\text{Reduction} \quad Cu^{++} + 2e = Cu \qquad (2.7)$$

(b) Electrode potentials indicate tendency for corrosion and deposition. Metals (ions) above hydrogen in the series are more readily reduced. Metals below hydrogen exhibit higher oxidation tendency.
(c) Use of hydrogen as a reducing agent under different pressures as well as that of oxygen (or other oxidants) under various partial pressures as an oxidizer for different metal ions and metals can be predicted.

In a bimetal combination, the metal with the nobler potential will act as cathode while the one with a relatively active potential will serve as anode. For example, in Fe–Zn couple, Fe will be cathodic to anodic Zn.

2.9 Limitation of EMF Series

- The series list only pure metals and not alloys and other composites.
- Rather than the thermodynamic electrode potentials for the various metal/metal ion concentrations, it is the corrosion potentials of metals and alloys in a defined corrosive environment which is of practical use and interest.

- EMF series predicts only the tendency for corrosion. There are metals in the EMF series such as chromium, aluminum and titanium which are listed as very active (negative potentials), but never the less do not corrode significantly due to passive oxide surface films. EMF series are silent on effect of environment on metal activity.

2.10 Galvanic Series [1–5]

Galvanic series overcome many of the drawbacks of the EMF series. Here the actually measured corrosion (rest) potentials of metals and alloys are listed in the order of their increasing—(or decreasing) activity in a defined environment (such as sea water) as shown in Table 2.2. It is of great practical importance since corrosion behavior of different alloys in combination can also be predicted.

Some alloys (such as 18-8 stainless steels) are shown as existing both in active and passive states. This could be seen in the light of the nature of surface passive films (whether stable or unstable (scratched surfaces).

The galvanic series would differ from environment to environment as well as to whether the media. (sea water, for example) is static or turbulent.

2.11 Electrochemical Aspects of Bimetallic (Galvanic) Corrosion [3, 6, 7]

Some basic conditions essential for bimetallic corrosion to occur are:
- Presence of continuous bridging between two metals (or alloys) through an electrolyte.
- Presence of concentration cells.
- Proper electrical contact and large potential difference among contacted metals (alloys).
- Sustained cathodic reaction at the nobler metal (alloys).

Major factors influencing galvanic corrosion in bimetallic couples include [6]

(a) Electrode potentials and electrode efficiency.
(b) Reactions such as metal dissolution, oxygen/hydrogen reduction.
(c) Metallurgical conditions such as composition, microstructure, alloy components, types of heat-treatment and mechanical working.
(d) Surface conditions like treatment, corrosion products and passive layers
(e) Electrolytic properties such as presence of various types of ions, pH, temperature, conductivity, aeration, flow rate and volume.

Table 2.2 Galvanic series

Platinum
Gold
Graphite
Silver
Hastelloy C
18-8 stainless steel (passive)
Chromium steel> 11% Cr (passive)
Inconel (passive)
Nickel (passive)
Monel
Bronzes
Copper
Brasses
Inconel (active)
Nickel (active)
Tin
Lead
Lead-tin solder
18-8 Mo stainless steel (active)
18-8 stainless steel (active)
Ni-resist
Chromium steel <11% Cr (active)
Cast iron
Steel or iron
2024 aluminium
Cadmium
Commercially pure aluminium
Zinc
Magnesium and its alloys

↑ Noble

↓ Active

(f) Environmental factors such as dry-wet cycles, water content, climatic and seasonal conditions as well as solar radiation.
(g) Geometrical aspects such as distance, positions, surface area, shape and orientations.

The compatibility of two different metals (or alloys) may be assessed through what is referred to as 'anodic index' [8], which is a measure of the electrochemical potential that is realized between the desirable metal and gold.

Typical anodic indices of some metals and alloys relative to most noble gold is given in Table 2.3.

To estimate the relative potential of a pair of metals or alloys it is only essential to subtract their anodic indices. For normal conditions, there should not be not more than 0.25 V difference.

The extent of galvanic effect need not always be related to differences in the electrode potentials as listed in the EMF series. The galvanic potentials measured in a given environment could be a better guide for assessing effect of potential differences on galvanic activity.

Titanium, aluminum and chromium possess highly active reversible potentials as projected in the EMF series, but occupy relatively nobler positions in the galvanic series. Galvanic corrosion of steel is higher when coupled to nickel and copper than when contacted with 304 stainless steel and Ti-6Al-4 V alloy. Reaction kinetics as well as nature of corrosion products may thus determine galvanic corrosion rates. Smaller quantities of alloying additions to a metal would not result in any significant shift in reversible potential, even though it could significantly influence the electrochemical kinetics. Multiphase microstructures can influence galvanic activity. Strange as it may seem, enhanced cathodic corrosion in a couple may happen as in the case of Zn–Al couple in saline solutions prematurely due to increased alkalinity near the electrode surface, when aluminum is not stable. Galvanic corrosion can occur in metal as well as multi-metal combinations. Presence of 'mixed metals' in piping is an example. In multi-contact situations, the most anodic metal would remain active whatever may be position of other metals, while the noblest metal would serve as cathode, irrespective of the different arrangements. However, the dissolution behavior of metals (and alloys) exhibiting intermediate potentials, depend on their relative positions in the multi-metal combination. Galvanic

Table 2.3 Anodic indices (V) [8]

Gold	0.00
Silver, high Ni–Cu alloys	0.15
Nickel, Ti and alloys	0.30
Copper, Ni–Cr alloys	0.35
2000 series wrought Al	0.75
Plain C and low alloy steels	0.85
Zinc	1.25
Magnesium and its alloys	1.75

corrosion can also occur in metal-nonmetallic material contacts such as metal-reinforced polymer matrix composites and metal-graphite composites.

Corrosion currents can be generated due to several reasons in metals and alloys.

- Presence of impurities.
- Grain boundaries and orientation.
- Differential temperatures/gradients.
- Surface roughness, surface product layers.
- Metallographic/micro-structural defects, inclusions, precipitates.
- Alloying elements and different phases.
- Differential stress/strain.

Bimetallic corrosion in the absence of physical contacts can also occur. Localized corrosion [7] on a metal can lead to formation of soluble corrosion products that can deposit through displacement reactions onto an active metal surface exposed to similar environmental conditions and form local anodic and cathodic cells. For example, in water heating systems, dissolved copper from the pipes can deposit on steel radiators, generating anodic and cathodic areas.

Iron corrosion products from steel fittings can flow over aluminum and deposit as cathodic magnetite.

Area, distance and geometric effects on bimetallic corrosion are very important with respect to design and operation of different industrial components. Highest galvanic corrosion rate in confined at the junction between two metals and severity of corrosion decreases with increased length (distance) as illustrated in Figs. 2.10 and 2.11.

Geometry and design of components would influence galvanic corrosion. Current does not flow around corners, as well.

When a current flows between anode and cathode in a corroding metal, the current will be the same across, independent of surface areas of each region. However, current density would differ depending on anodic and cathodic surface area ratios. The larger the cathode, compared to the anode, higher oxygen reduction (or similar cathodic reactions) can occur and hence the galvanic currents are enhanced. The effect of the ratio of the anodic to cathodic areas thus becomes a

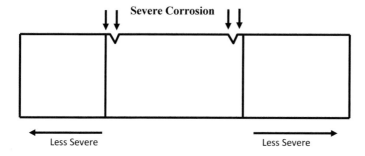

Fig. 2.10 Model showing severe galvanic corrosion at junctions-distance effect

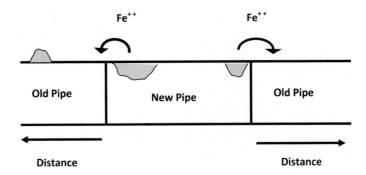

Fig. 2.11 Model showing distance effect in galvanic corrosion

significant factor controlling galvanic corrosion rates. A small anode area in contact with larger cathodic area results in serious bimetallic anodic corrosion due to higher anodic current densities on smaller anodes (see Fig. 2.12). As a general rule of principle, anode area should be larger than that of the cathode to minimize galvanic corrosion. For corrosion protection, the cathode regions in a component should be painted (coated) based on area-effect, discussed above. If the paint is damaged, then a smaller cathode in contact with relatively larger anode would be exposed, minimizing corrosion rates. On the other hand, if the anode is painted, paint damage would result in the formation of a smaller anode in contact with larger cathode, leading to enhanced corrosion rates. For example, in a carbon steel (anode) structural component in contact with stainless steel (cathode), surface coating of only the carbon steel could lead to disastrous corrosion due to unfavorable area effect [2–5].

In this regard, superiority of galvanized steel components need to be stressed. A uniformly zinc coated steel surface when exposed to a corrosive environment will be protected under all conditions of coating damage. Even if large portions of zinc coatings are abraded away, the base steel will still be protected due to favorable area effect! (see Fig. 2.13).

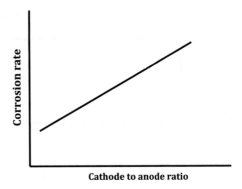

Fig. 2.12 Variation of corrosion rate with increasing catholic area for a fixed small anodic area

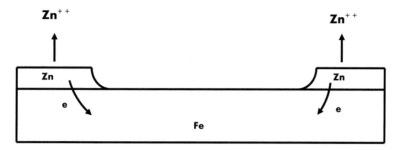

Fig. 2.13 Sacrificial zinc corrosion in contact with iron (Galvanic coating)

Fig. 2.14 Model showing area effect in galvanic corrosion. **a** Iron rivets on copper plates. **b** Copper rivets on iron sheets

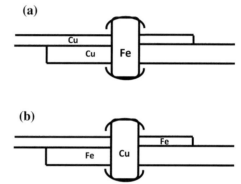

Steel rivets on a copper bar exposed to sea water is yet another example to unfavorable area effect compared to copper rivets on a steel bar (see Fig. 2.14).

Yet another significant observation is polarity reversal under certain environmental conditions with time. Some examples are illustrated below [1–7].

Tinning is used to protect steel containers. Internally tin-coated steel cans are used to preserve vegetable and many types of fruit juices, Tin in nobler to iron in the EMF series. However, tin can dissolve to form Sn^{++} due to the presence of organic acids from the stored vegetable and fruit juices. Stannous ions can form tin complexes with organic acids, leading to reversal of polarity of the

$$Sn = Sn^{++} + 2e \text{ reaction} \qquad (2.28)$$

Reversible Sn/Sn^{++} potential could shift to more active values. $\left[\dfrac{Sn^{++}}{Fe^{++}}\right]$ ratios corresponding to onset of polarity reversal can be estimated from the reaction,

$$Fe^{++} + Sn = Fe + Sn^{++} \text{ when } E_{Cell=0} \qquad (2.29)$$

$\left[\dfrac{Sn^{++}}{Fe^{++}}\right]$ must be less than 5×10^{-11} for tin to become more active than iron [5].

Change of surface condition of at least one among the metallic couples can cause polarity reversal. For the zinc–steel couple, the change in the zinc electrode potential is mainly responsible for polarity reversal since the iron potential does not significantly change with time in hot water. Passivation of the zinc surface in presence of oxygen in water lead to cathodic depolarization.

Polarity reversal can occur in aluminum-steel couples as well (used in cathodic protection). In presence of carbonate/bicarbonate ions, aluminum surface passivation may be promoted, shifting potential to nobler values. Polarity reversals as in the zinc-steel and aluminum-steel couples result in loss of cathodic protection of the steel component.

In the light of above bimetallic corrosion fundamentals, the following prevention or control methods can be suggested [2–6]:

(a) Selection of metals or alloy combinations as close together as possible in the galvanic series.
(b) Due importance to be given to surface area ratio effect and contacts to be avoided where the area of the active metal (alloy) is smaller. In case of fasteners, always prefer nobler metal components. Design of anodic parts in an assembly should take into consideration area and geometry.
(c) Wherever possible, dissimilar metal contacts need to be insulated from each other.
(d) Coatings or painting need caution. Do not paint less noble contact portions without also coating the nobler one. If only one contact surface need to be coated, the nobler surface to be coated.
(e) Metals need to be kept as far as possible (distance effect).
(f) Inhibitors could reduce corrosiveness. Cathodic protection can be used wherever suitable.

2.12 Potential-pH Diagrams [1–5]

Eh (Electrode potential with reference to standard hydrogen electrode) and pH ($-\log a_{H+}$) are the major environmental parameters influencing aqueous corrosion of metals. Electrochemical equilibrium diagrams can be constructed based on thermodynamic principles to predict corrosion and protection of different metals in an aqueous medium as a function of Eh and pH for different oxidation/reduction reactions and pH levels. Basic Eh-pH diagram as a two dimensional representation in an aqueous phase consists of four coordinates representing oxidizing and reducing regions across acidic and pH levels as illustrated in Fig. 2.15. Increasing of Eh towards more positive values indicate enhanced oxidizing environments compared to lower, less positive Eh values, which denote shift towards reducing environment.

Upper and lower stability limits of water can be established for the Eh-pH diagram based on the following redox reactions (Fig. 2.16):

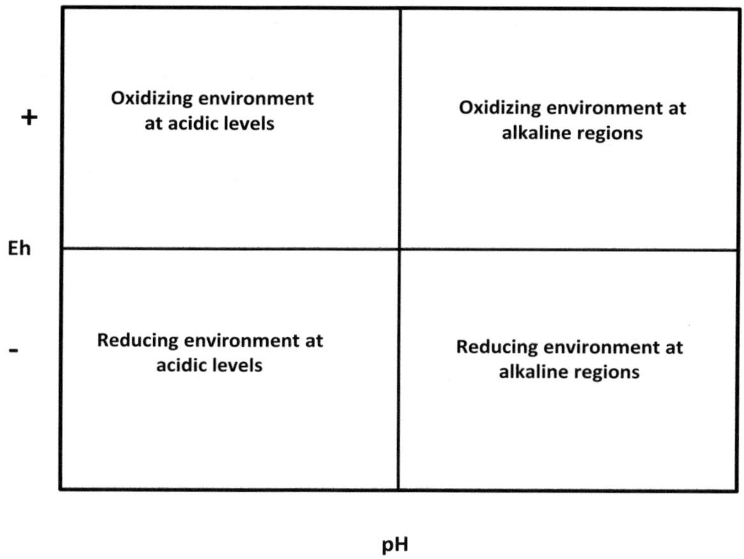

2 Principles of Corrosion Processes

Oxidizing environment at acidic levels	Oxidizing environment at alkaline regions
Reducing environment at acidic levels	Reducing environment at alkaline regions

Fig. 2.15 Eh-pH basic coordinates

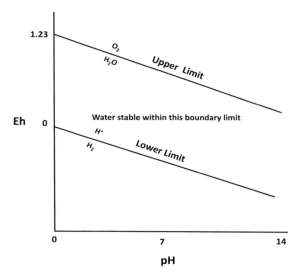

Fig. 2.16 Stability region for water in the Eh-pH diagram

(a) $$O_2 + 4H^+ + 4e = 2H_2O \quad E^0 = +1.23V \quad (2.30)$$

$$Eh = 1.23 - 0.059\,pH \,(\text{at}\,pO_2 = 1)$$

(b) $$2H^+ + 2e = H_2 \quad E^0 = 0.00V \quad (2.4)$$

$$Eh = 0 - 0.059\,pH \,(\text{at}\,p_{H2} = 1)$$

The stability limits can shift depending on oxygen and hydrogen partial pressures. For complete representation of the above water stability diagram, reactions involving oxygen reduction to form H_2O_2 which further reduces to H_2O need to be considered under neutral and alkaline conditions. Reactions such as

$$O_2 + 2H_2O + 4e = 4OH^- \quad (2.21)$$

$$2H_2O + 2e = H_2 + 2OH^- \quad (2.31)$$

are also likely. For various metal-water-oxygen systems, stability regions for oxidized and reduced species fall within the above marked boundaries.

Three types of equilibrium states can be realized based on the following types of reactions:

(a) Depending only on Eh, but independent of pH (Horizontal line to the X-axis).
(b) Dependent only on pH, but independent of Eh (Vertical to the X-axis).
(c) Dependent on both Eh and pH (line with defined slope).

Various types of reactions are illustrated in Fig. 2.17.

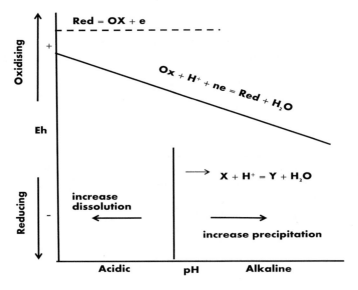

Fig. 2.17 Possible reactions in an Eh-pH diagram

2 Principles of Corrosion Processes

Eh-pH diagrams are drawn for specific ion activities and partial pressures of gases at room temperature from thermodynamic data. For the Fe–H_2O–O_2 system, major thermodynamically feasible reactions are illustrated below, assuming Fe^{++}, Fe^{+++} activities at 1M and gaseous partial pressures as one at room temperature [1].

1.
$$Fe = Fe^{++} + 2e \quad E^0 = -0.44V \tag{2.27}$$
(Reaction dependent only on Eh, independent of pH)

$$Eh = -0.44 + \frac{0.059}{2} \log[Fe^{++}] \tag{2.32}$$

$$\text{For } a_{Fe^{++}} = 1, Eh = -0.44V \tag{2.33}$$

2.
$$Fe^{++} + 2H_2O = Fe(OH)_2 + 2H^+ \tag{2.34}$$
(Reaction dependent only on pH, independent of Eh)

$$\Delta F^0 = -RT \ln K$$
$$= -1.364 \log K \tag{2.35}$$

$$2pH + \log[Fe^{++}] = 13.28 \tag{2.36}$$

$$\text{Log}[Fe^{++}] = 13.28 - 2pH \tag{2.37}$$

$$pH = 6.64 \tag{2.38}$$

3.
$$Fe^{++} = Fe^{+++} + e \quad E^0 = +0.771V \tag{2.39}$$
(Eh dependent, but independent of pH)

$$Eh = 0.771 + 0.059 \log \left[\frac{Fe^{+++}}{Fe^{++}}\right] \tag{2.40}$$

$$Eh = 0.771V \tag{2.41}$$

4.
$$Fe^{+++} + 3H_2O = Fe(OH)_3 + 3H^+ \tag{2.42}$$
(reaction dependent only on pH)

$$\text{Log}[Fe^{+++}] = 4.81 - 3pH \tag{2.43}$$

$$pH = 1.6 \tag{2.44}$$

5.
$$Fe + 2H_2O = Fe(OH)_2 + 2H^+ + 2e \quad (2.45)$$
(dependent both on Eh and pH)

$$E^0 = -\frac{2.19}{23.06X2} = -0.05V \quad (2.46)$$

$$Eh = -0.05 + \frac{0.055}{2}\log[H^+]^2 \quad (2.47)$$

$$Eh = -0.05 - 0.059 pH \quad (2.48)$$

6.
$$Fe(OH)_2 + H_2O = Fe(OH)_3 + H^+ + e \quad (2.49)$$

$$Eh = 0.27 - 0.059\,pH \quad (2.50)$$

7.
$$Fe^{++} + 3H_2O = Fe(OH)_3 + 3H^+ + e \quad (2.51)$$

$$Eh = 1.057 - 0.177 pH - 0.059\log[Fe^{++}] \quad (2.52)$$

The above seven reactions are then drawn on an Eh-pH diagram. Stability regions for Fe, Fe^{++}, Fe^{+++}, $Fe(OH)_2$ and $Fe(OH)_3$ phases are marked as shown in Fig. 2.18.

The corrosion diagram corresponding to the $Fe-H_2O-O_2$ equilibrium is shown in Fig. 2.19.

With changes in the ionic concentrations of dissolved iron and partial pressures of oxygen and hydrogen, the phase boundaries would shift.

Regions of corrosion for iron (steels) are not only confined to the acidic region, but also exist in the high alkaline region (beyond a pH of about 12) where dissolution of iron as $HFeO_2^-$ species can occur. Regions of immunity (where metallic iron is stable thermodynamically) and passivation (stability phases for iron oxides which form a protective passive layer) are shown. From the diagram, suitable Eh-pH regions for cathodic and anodic protection could be selected. Similar Eh-pH and corrosion diagrams can be drawn for different metal—H_2O-O_2 systems to understand their corrosion behavior. Representative corrosion diagrams for aluminum, zinc, magnesium, nickel, copper and titanium are shown in Fig. 2.20.

There are some limitations to thermodynamically constructed Eh-pH diagrams.

- The diagrams are thermodynamically derived for room temperature conditions. Corrosion behaviour at higher temperatures cannot be predicted using this diagram.
- Only thermodynamic amenability to corrosion and protection are predicted. Corrosion kinetics cannot be assessed.

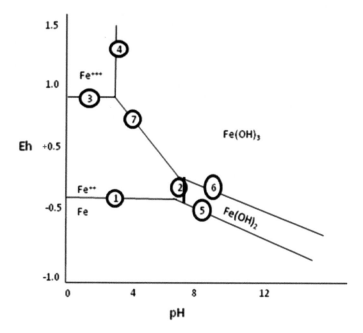

Fig. 2.18 Eh-pH diagram for the Fe–H_2O–O_2 system showing seven reaction boundaries

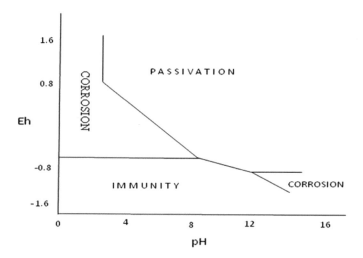

Fig. 2.19 Corrosion diagram for iron

- No consideration for added ions and effect of metal complexation is given.
- Only pure metals are considered. Effects of alloying and metallographic phases and heterogeneities are not considered.

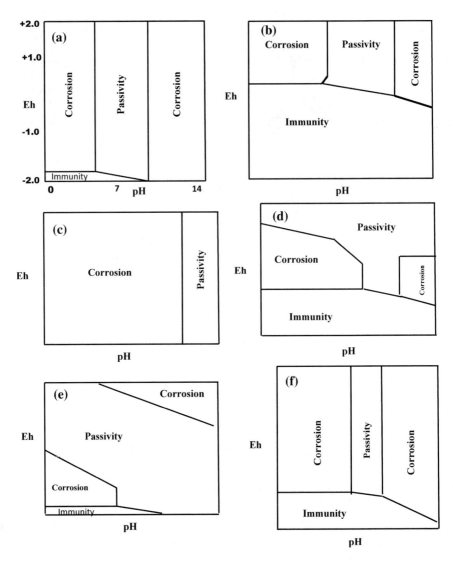

Fig. 2.20 Corrosion diagrams of **a** Aluminum, **b** Copper, **c** Magnesium, **d** Nickel, **e** Titanium, **f** Zinc

Modified Eh-pH diagrams for various metals need to be prepared taking into consideration the effect of alloying additions, presence of complexing agents and higher temperatures. Combined diagrams taking into consideration two or more metals simultaneously present would be more useful. For example, effect of chromium addition to iron to increase corrosion resistance can be represented in a combined Fe–Cr diagram. Similarly Cu–Zn, Cu–Sn and aluminum and magnesium alloy diagrams can be constructed.

Computer programs and soft-ware kits are now available to construct Eh-pH diagrams for various metal systems under different environmental conditions.

2.13 Electrochemical Kinetics [1–5, 9]

When a metal is contacted with an electrolyte containing its ions, either oxidation (loss of electrons) or reduction (gain of electrons) can occur. Equilibrium conditions existing across the metal-solution interface will determine the nature of the reactions. Electron transfer across interface will take place until thermodynamic equilibrium is reached. At equilibrium, the net current is zero while the oxidation-reduction rates are equal. The reversible electrode potential corresponds to the potential at equilibrium while the net equivalent current across the interface when there is no external current is supplied is referred to as the exchange current (i_o). When a metal is at its equilibrium potential in a solution, rates of oxidation and reduction are equal (not zero)

$$r_1 = r_2 = \frac{i_o a}{nF} \qquad (2.53)$$

where r_1 and r_2 are rates of forward and reverse reactions and i_o is the exchange current density.

When a net current flows through a corrosion cell, the difference between the measured potential and the reversible half-cell potential ($\Delta E = E - E_{eq}$) is called the over-potential which is a measure of the departure (or deviation) of the potential from its equilibrium value. An electrode is not in equilibrium when a net current flows from or to its surface and to enable the current flow, the electrode potential should shift off its equilibrium value.

Polarization essentially results from a slow step in an electrode process. Considering transport processes at a metal-solution interface, diffusion of electroactive ions from the bulk to the interface and their subsequent interaction at the electrode surface resulting in charge transfer can be considered to understand the type of polarization.

A slow step in diffusion transport of reactants to and from the electrode interface results in concentration polarization, while a slow step in the charge transfer step results in activation polarization.

$$\eta_{Total} = \eta_{concentration} + \eta_{activation}$$

Depending on as to whether the polarization is significantly at the anode or cathode or at both anode and cathode, anodic, cathodic and mixed control electrode processes can be realized. Anodic, cathodic and mixed controls are represented in polarization diagrams shown in Fig. 2.21.

Exchange current densities can only be determined experimentally.

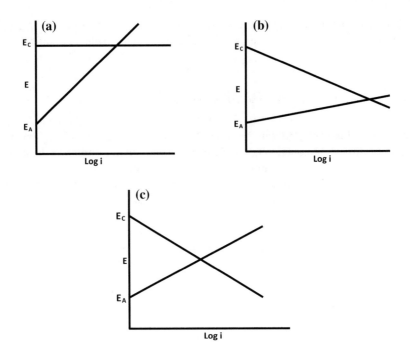

Fig. 2.21 Anodic (**a**), Cathodic (**b**) and Mixed (**c**) Control

A kinetic expression for exchange current density can be derived as

$$i_o = nFAK_s(c_{ox})^{1-\alpha}(c_{red})^{\alpha} \tag{2.54}$$

where

K_s is a rate constant.
A = surface area.
α = transfer coefficient.

Exchange current depends on the following

- Nature of the redox reaction.
- Electrode surface and composition.
- Temperature.
- Concentration of redox species.

Exchange current densities (amp/cm^2) for various metals for the hydrogen reduction reaction are given below [1–5]:

2 Principles of Corrosion Processes

Pb, Hg	10^{-12} to 10^{-13}
Zn	10^{-10} to 10^{-11}
Ag, Cu	10^{-7}
Fe, Au	10^{-6}
Pd, Rh	10^{-4}
Pt	10^{-2}

It becomes 10^{12} times easier for hydrogen to evolve on a platinum cathode compared to lead and mercury surfaces. Thus, the magnitude of exchange current density indicates the reversibility or irreversibility of a redox reaction at an electrode surface.

Electrochemical kinetics of corrosion can be derived in terms of activation and concentration polarization for anodic and cathodic reactions.

From the rate theory equations, relationships between applied current density, exchange current density and activation over potentials can be derived.

A general relationship for the polarization of an electrode for a specific redox reaction is given by the Butler-Volmer equation [2].

$$i_a = i_o \exp\left[\frac{\propto nF\eta}{RT}\right] \tag{2.55}$$

$$i_c = i_o \exp\left[\frac{-(1-\propto)nF\eta}{RT}\right] \tag{2.56}$$

$$I_{applied} = i_0 \left[\exp\left(\frac{\propto nF\eta}{RT}\right) - \exp\left(\frac{-(1-\propto)nF\eta}{RT}\right)\right] \tag{2.57}$$

where

$I_{applied}$ = net reaction current
$\propto$ = charge transfer or symmetry coefficient for the anodic/cathodic reaction

The factors $\propto$ and $(1-\propto)$ are the fractions of over-potential (η) taken by discharge and ionization reactions (forward and backward)

n = number of electrons
F = Faraday constant
T = Absolute temperature.

For high over-potentials, the above equation can be simplified as the Tafel equation

$$\eta_{act(anode)} = a + b_a \log\frac{i_a}{i_o} \tag{2.58}$$

$$\eta_{act(cathode)} = a - b_c \log \frac{i_c}{i_o} \qquad (2.59)$$

$$\eta_{act} = \pm b \log \frac{i}{i_o} \qquad (2.60)$$

$\propto = 0.5$ for b_a and b_c of 0.12 V
Anodic and cathodic slopes can be estimated as

$$b_c = -2.303 \frac{RT}{\propto nF} \qquad (2.61)$$

$$b_a = 2.303 \frac{RT}{(1-\propto)nF} \qquad (2.62)$$

Tafel equation is applicable to each electrode half reaction separately.

At low over-potentials, the dependence of current on over-potential is linear and the linear region is referred to as 'polarization resistance'.

As per the Stern-Geary equation, applicable to the linear region at lower over potentials,

$$i_{Corr} = \frac{B}{R_p} \text{ and } B = \frac{b_a \cdot b_c}{2.3(b_a + b_c)} \qquad (2.63)$$

where R_p = polarization resistance = $\frac{\Delta E}{\Delta I}$.

Tafel and stern-Geary equations are used to estimate corrosion currents (corrosion rates) by the Tafel plots and linear polarization methods.

Similarly, from Faraday's laws and Fick's laws of diffusion, a relationship for concentration polarization with reference to limiting (diffusion) current and cathodic current can be derived.

$$\eta_{Conc} = \frac{RT}{nF} \ln\left[1 - \frac{i_C}{i_L}\right] \qquad (2.64)$$

where i_L = limiting (diffusion) current density.

Total cathode polarization consisting of activation and concentration over-potentials can be written as

$$\eta_{total} = -\beta \log \frac{i_C}{i_O} + 2.3 \frac{RT}{nF} \log\left[1 - \frac{i_c}{i_L}\right] \qquad (2.65)$$

Figure 2.22 illustrates corrosion of a metal where cathodic reaction is under activation and concentration polarization.

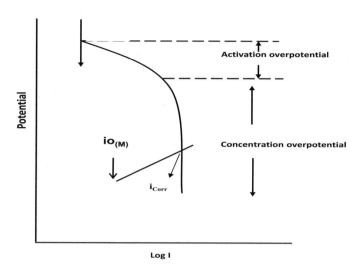

Fig. 2.22 Polarization diagram showing activation and concentration over-potentials [2, 4]

2.14 Theory of Mixed Potentials [1–5, 10]

Electrochemical fundamentals governing mixed potential theory are based on the following concepts:

- Principle of charge conservation: There cannot be any net accumulation of charge during an electrochemical reaction. Total rate of oxidation must be equal to total rate of reduction. i.e. Sum of anodic oxidation currents must be equal to sum of cathodic reduction currents.
- Any electrochemical reaction can be divided into two or more partial oxidation and reduction reactions.

    Consider a general anodic reaction,

$$M = M^{++} + 2e \tag{2.1}$$

Possible cathodic reactions depending on environment are:

(a) Hydrogen reduction from acid or neutral/alkaline solution.

$$2H^+ + 2e = H_2 \, (\text{acid}) \tag{2.4}$$

$$2H_2O + 2e = H_2 + 2OH^- \, (\text{neutral or alkaline}) \tag{2.31}$$

(b) Oxygen reduction in acid or neutral solution.

$$O_2 + 4H^+ + 4e = 2H_2O \text{(acid)} \tag{2.30}$$

$$O_2 + 2H_2O + 4e = 4OH^- \text{(neutral)} \tag{2.21}$$

(c) Reduction of other added oxidizers such as ferric ions

$$Fe^{+++} + e = Fe^{++} \tag{2.39}$$

As per the mixed potential theory using the zero current criterion, $\sum i_a = \sum i_c$

$$i_a^M = i_c^{H(M)} + i_{c\,2}^{O\,(M)} + i_c^{Fe^{+++}(M)} \tag{2.66}$$

At equilibrium, the total anodic oxidation rate is equal to total cathodic reduction rate.

Assuming corrosion of an divalent metal, M in an acid solution,

$$M = M^{++} + 2e \text{(anodic reaction)} \tag{2.1}$$

$$\underline{2H^+ + 2e = H_2 \text{(cathodic reaction)}} \tag{2.4}$$

$$M + 2H^+ = M^{++} + H_2 \text{ (net reaction)} \tag{2.67}$$

The metal M corrodes with the evolution of hydrogen. The two half-reactions as indicated above cannot coexist as separate entities on the same metal surface. Each reaction has its own half-cell electrode potential and exchange current density.

Due to polarization, potentials shift in anodic and cathodic directions to an intermediate value (between the two half-cell potentials).

Since such a polarized potential is a combination of the two half-cell potentials, it is referred to as MIXED POTENTIAL. The electrode potential at steady state for a freely corroding metal is referred to as corrosion potential (E_{Corr}). See Fig. 2.23.

At E_{corr}, rates of anodic and cathodic reactions are equal.

$$i_c = i_a = i_{corr}, \text{(at } E_{corr}) \tag{2.68}$$

i_{corr} is the corrosion rate of the metal and also the rate of hydrogen liberation at the metal surface (H^+ oxidizes the metal).

Kinetic parameters such as the exchange current density for the redox reaction at a metal surface need to be considered to understand the reversibility of a redox reaction at metal surfaces.

For example, in dilute hydrochloric acid solutions, zinc dissolution is expected to be higher than that of iron from a thermodynamic view point ($E^0_{Zn/Zn^{++}} = -0.76$ V compared to E^0 for $Fe|Fe^{++} = -0.44$ V). However, from a kinetic view point, the

Fig. 2.23 Polarization of anodic and cathodic reactions resulting in a mixed potential

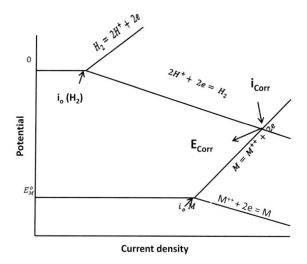

corrosion rate of iron will however be higher than that of pure zinc, due to differences in their exchange current densities for hydrogen liberation reaction. Exchange current density for hydrogen reduction on zinc is lower than that on iron.

Engineering systems are heterogeneous and complex. The zero current criterion in such multi-electrode systems in a corrosive environment becomes all the more relevant.

Consider two electrodes X and Y with one reduction reaction in an acid solution [11].

$$i_a^X + i_a^Y = i_c^{H(X)} + i_c^{H(Y)} \qquad (2.69)$$

Relative areas of the anode and cathode are important in the prediction of anodic corrosion rates and current density (current/unit area) need to be considered.

The driving force for corrosion is enhanced on addition of a strong oxidizer such as ferric ions to the acid solution. The corrosion potential E_{corr}, is shifted to more noble direction with increasing corrosion rate of the metal. Hydrogen reduction rate is correspondingly decreased due to the added oxidizer. See Fig. 2.24.

$$\begin{aligned} I_{corr} &\text{(Corrosion rate on addition of ferric ion)} \\ &= i_C^{Fe^{+++}/Fe^{++}} + i_C^{H^+/H_2} \text{(at steady state)} \end{aligned} \qquad (2.70)$$

It may however be borne in mind that the above shown effect of an added oxidizer will be significant only if the exchange current density for the added oxidizer reduction is higher than that for hydrogen reduction. If the exchange current for the added oxidizer reduction is lower, no increase in corrosion rate of metal would be observed.

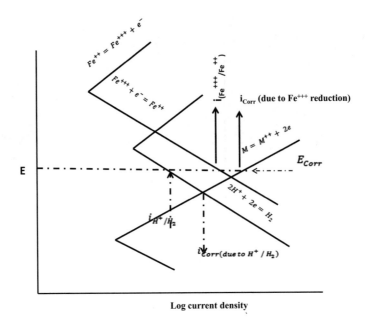

Fig. 2.24 Mixed potential diagram illustrating effect of addition of ferric ions to acid solution on the corrosion rate of a metal [9, 10]

Let us examine the corrosion behavior of active-noble metal couples under the mixed potential theory.

2.15 Platinum-Iron Couple in Acid Solution [11]

$$i_a^{Fe} = i_c^{H(Pt)} \frac{A(Pt)}{A(Fe)} + i_c^{H(Fe)} \qquad (2.71)$$

where i_a and i_c represent anodic and cathodic current densities and A is relative surface area. Corrosion rate of platinum in the couple is negligible, while that of iron is enhanced due to very high exchange current density for hydrogen reduction on platinum. Surface area effect also needs to be considered. Platinum provides additional cathode surface for efficient hydrogen reduction.

The corrosion rate of an active metal (Fe, Zn) depends on the nature of the coupled noble metal in relation to the exchange current density for the reduction reaction. pH of the medium and nature of the oxidizer will also influence rate of anodic oxidation. If the Fe–Pt couple (having higher cathode to anode surface area ratio) is exposed to neutral pH solution where oxygen reduction is the cathodic reaction (instead of H_2), the expected effect of noble metal (Pt) would be not so significant since the exchange current densities for oxygen reduction on both the surfaces are nearly the same. Also, if lead metal is coupled with iron instead of platinum in acid solution, the effect of coupling on anodic oxidation of iron would be rather negligible, since the exchange current density for hydrogen reduction on lead is very much lower than that on iron.

2.16 Iron-Zinc Couple [11]

Corrosion behavior of the iron-zinc couple can be examined with respect to role of zinc as a sacrificial anode

$$i_a^{Fe} = i_c^{H(Zn)} \frac{A(Zn)}{A(Fe)} + i_c^{H(Fe)} - i_a^{Zn} \frac{A(Zn)}{A(Fe)} \tag{2.72}$$

$$i_a^{Zn} \gg i_c^{H(Zn)} \tag{2.73}$$

Enhanced zinc oxidation rate would decrease the rate of iron oxidation and the effect zinc depends on the ratio of $\frac{i_c^{H(Zn)}}{i_a^{Zn}}$ and iron will be protected as long as this ratio is smaller.

The relative areas of the two metals in a couple influence its galvanic corrosion rate. Increasing cathodic surface areas for a fixed anode area will increase the anodic corrosion rate.

While E_{corr} can be directly measured, i_{corr} need to be determined by polarizing the electrode from the corrosion potential.

The entire corroding metal is made either anodic or cathodic in an electrolytic cell through application of an external potential (or current) and steady state conditions deduced from the polarized condition [2].

$$\eta_{act} = b_a \log \frac{i_{net}}{i_{corr}} \text{ at higher anodic over-potentials} \tag{2.74}$$

For cathodic reaction,

$$\eta_{act} = b_c \log \frac{i_{net}}{i_{Corr}} \tag{2.75}$$

$$i = i_{corr} \text{ when } \eta = 0 \tag{2.76}$$

Anodic and cathodic Tafel lines can be extrapolated back to E_{corr} to get i_{corr} (corrosion rate).

When the cathodic reaction is diffusion controlled as in the case with oxygen reduction at neutral pH,

$$O_2 + 2H_2O + 4e = 4OH^- \quad (2.21)$$

solution velocity influences corrosion rate (unlike in activation control). Corrosion rate of normal metals initially increases with solution velocity up to a certain value, there after becoming independent of velocity.

2.17 Determination of Corrosion Rates [2]

Electrochemical reaction rate can be measured in terms of rate of electron transport from or to the electrode interface.

As per Faraday's law,

$$M = \frac{It.a}{nF} \quad (2.77)$$

where

M = mass reacted
a = atomic weight
I = current, in amperes
t = time
n = number of exchanged equivalents
F = Faraday constant

Proportionality between mass loss per unit area per time and current density can be given as

Corrosion rate,

$$r = \frac{M}{A.t} = \frac{ia}{nF} \quad (2.78)$$

where

$$i = \text{current density} \left(\frac{i}{A}\right)$$

$$\text{Penetration rate} = 0.129 \frac{a.i}{nd} \text{ in mpy(mils per year)} \quad (2.79)$$

where

d = density of metal/alloy in g/cm^3
i = $\mu A/cm^2$ and 0.129 is the proportionality constant

For expressing penetration rate in other units such as mm/year or μm/year appropriate proportionality constants need to be used.

As an example, the relationship between penetration rate in mpy and current density of 1μA/cm² for iron can be calculated as

$$0.129 \left[\frac{55.8 \times 1}{7.9 \times 2} \right] \approx 0.5 mpy = 1 \: \mu A/cm^2$$

To determine the corrosion penetration rate for alloys, respective metallic compositions need to be considered and the equivalent weight determined in terms of sums of fractions of equivalents of all alloying elements.

$$\text{Total equivalents } N_t = \Sigma \frac{f_i n_i}{a_i} \tag{2.80}$$

where

f_i = mean fraction of the element in the alloy
n_i = exchanged electrons
a_i = respective atomic weights

From corrosion current densities derived from polarization plots, equivalent penetration rates in mass loss per unit area per time can be estimated.

Typical Tafel extrapolation and linear polarization plots for determination of corrosion currents and other electrochemical parameters are illustrated in Figs. 2.25 and 2.26.

Fig. 2.25 Estimation of Tafel parameters

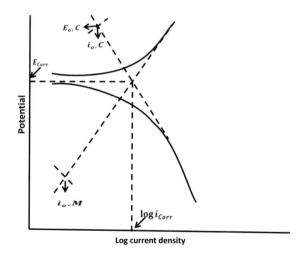

Fig. 2.26 Linear polarization resistance method

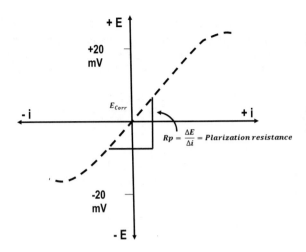

2.18 Electrochemical Aspects of Passivity [1–5, 11, 12]

In the Eh-pH diagrams, stability regions of immunity, corrosion and passivity are indicated for various metals in aqueous solutions. Properties such as immunity and passivity can be effectively used to protect metals form corrosion.

Electrochemical aspects of passivity and methods of corrosion protection utilizing active- passive transition are illustrated below:

Passivity can be defined as follows [5]

(a) A metal active in the EMF series or an alloy constituted of such metals can be defined as passive when its electrochemical behavior or activity becomes same or similar to that of a less active (nobler) metal. Eg: Titanium, Chromium, Stainless steel.

Passive metals such as titanium, chromium, aluminum and alloys such as stainless steel are corrosion resistant due to the formation of very thin, adherent and protective oxidized surface films in the corrosive environment.

(b) A passive metal or alloy effectively resists corrosion in an environment even if there exists a negative free energy change for its conversion from a metallic state to appropriate corrosion products.

Eg: Lead in sulfuric acid, Iron in inhibitor-containing pickling acid.

There are several theories put forward to explain passive behavior of metals and alloys such as oxide film and adsorption theories and appropriate models proposed.

Schematic anodic polarization behavior of a metal exhibiting active-passive transition is illustrated in Fig. 2.27 [1, 2].

Active, passive and transpassive regions are clearly shown. From the anodic polarization curve, the following electrochemical parameters can be deduced.

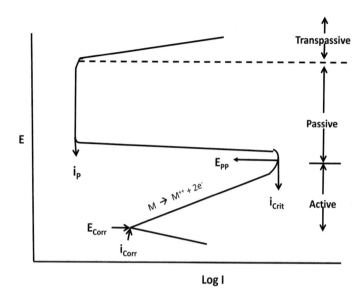

Fig. 2.27 Schematic anodic polarization curve illustrating active-passive transition [1–5]

i_{crit} Critical anodic current density (minimum anodic current density essential to initiate passivation-sudden decrease in corrosion rate from the initial active region)

E_{pp} Primary passive potential (Above this potential, the passive film becomes stable and corrosion rate decreases drastically to the lower value)

E_F Flade potential (on interruption of anodic polarization, decay in passivity with time occurs through potential changes in stages. The noble potential exhibited just before rapid potential decay was found to be more noble, the more acid the solution and referred to as Flade potential [5]. See Fig. 2.28.

i_p Passive current density is the minimum current density required to maintain stable passivity.

E_{pit} Pitting potential is the potential at which there is sudden increase in current density due to breakdown of passive film. At higher potentials beyond the stable passive region, the passive film is disturbed and breaks down, leading to subsequent increase in anodic corrosion rate in the transpassive region

It is possible to determine, the passive corrosion rate, passive potential region, relative stability of the passive state as well as the essential electrochemical conditions necessary to achieve spontaneous passivity.

Enhancement of temperature and acidity tends to increase the critical anodic current density for passivation. Presence of chlorides is detrimental to passivity.

Mixed potential behavior of active–passive metals and alloys can be understood when cathodic reduction processes are superimposed on the anodic polarization curve. As shown in Fig. 2.29, three different activation controlled reduction

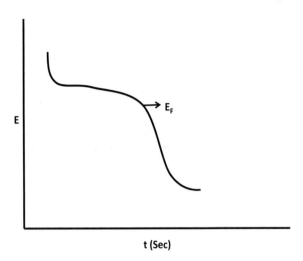

Fig. 2.28 Decay of passivity on interruption of anodic polarization indicating Flade potential (E_F) [5]

processes having varying exchange current densities are superimposed on the anodic polarization curve [2, 12].

Three different conditions are possible, namely [12]

- Case I
  Only one stable potential at C where the mixed potential theory is satisfied. The metal corrodes having i_{corr} and E_{corr} corresponding to point C
  Eg: Fe in dilute H_2SO_4, Ti in dilute H_2SO_4/HCl.

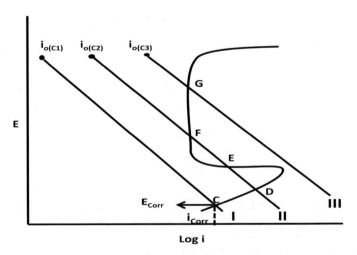

Fig. 2.29 Stability of passivity under different activation—controlled cathodic processes [12]

- Case II
  Three points of intersection D, E and F where rate of oxidation is equal to rate of reduction. Point E is not in stable state, while D is in active region (high corrosion rate) and F in passive state (lowest corrosion rate). This system can exist in active and passive stable states. (Borderline passivity)
  Eg: Cr in dilute HCl or H_2SO_4. Stainless steel in H_2SO_4 (containing oxidizers).
- Case III
  Spontaneous passivation with only stable potential at G in the passive region. This case is the most desirable.
  Eg: Cr—noble metal alloys in H_2SO_4 or HCl.
  Ti—noble metal alloys in dilute H_2SO_4.
  18-8 stainless steel in acid (containing ferric ions).

Achievement of condition as in case III is essential for the development of corrosion resistant alloys. Borderline passivity need to be avoided in which both active and passive states remain stable. At higher oxidizing conditions, passive films break down and transpassivity sets in, leading to initiation of localized corrosion through pitting.

Total cathodic partial current density at E_{pp} should be equal to or greater than i_{crit} to achieve spontaneous passivation.

The electrochemical criterion to achieve spontaneous passivation can be stated as [12]:

$$\text{Passivity Index (PI)} = \frac{i_c(atE_{pp})}{i_{crit}} \qquad (2.81)$$

For PI $\geq$ 1, Spontaneous passivation occurs and for PI < 1, no spontaneous passivation occurs, even though as in case II, a stable passive region may exist along with another corrosion region (Borderline passivity).

Spontaneous passivation depends on

(a) Passivation potential
(b) Reversible potential of the oxidant
(c) Passivation current density
(d) Cathodic partial current density of the oxidant at passivation potential.

To achieve stable spontaneous passivation, the following electrochemical conditions need to be satisfied.

- Passive potential to be lower than reversible potential of the oxidant.
- Passive current density to be less than cathodic reduction current density.

The choice of a metal or alloy based on passive corrosion resistance depends on the following factors [12]:

(a) To achieve passive behavior where cathodic reduction is activation controlled, a metal or alloy with an active E_{pp} is superior.

(b) If the reduction process is diffusion controlled, a metal or alloy having a small i_{crit} will passivate faster.

For development of corrosion-resistant alloys based on passivity criterion, two approaches then become possible.

(a) Enhancing ease of passivation by reducing i_{crit} or keeping E_{pp} at more active values.
Anodic dissolution behavior of the metal/alloy can be changed by appropriate alloy addition in order to decrease i_{crit}.
Eg: Titanium, chromium
(alloying additions such as molybdenum, nickel, tantalum and columbium beneficial)
(b) Increase cathodic reduction rates in order to achieve spontaneous passivation—coupling/alloying with noble metals having high exchange current densities for reduction processes.

For example metals with active E_{pp} such as titanium and chromium and alloys containing these metals which possess high exchange current densities for hydrogen reduction can undergo spontaneous passivation.

Rather unusual effects are observed in galvanic contacts between active-passive metals and noble metals such as platinum [4]. For example, titanium when coupled to platinum in an acid solution in the absence of oxidizers exhibits spontaneous passivation. Titanium can exist in passive condition at potentials more active than the reversible hydrogen reduction potentials. The exchange current density for hydrogen reduction on platinum is very high and the reduction current is larger than the critical anodic current density for titanium passivation, under the above conditions. On the other hand, if the passive range for the active-passive metal begins at nobler potentials than the reversible hydrogen potential, the corrosion rate of 'active' titanium will increase when coupled to 'noble' platinum.

Effect of addition of oxidizer (ferric, chromate) on the electrochemical behavior of active-passive alloys can be compared with those of normal metals. Corrosion rate of an active-passive alloy initially increases with oxidizer additions, while still in active state. However, after passive state is reached, the corrosion rate steeply decreases to a lower value and essentially remains at this low corrosion rate thereafter. Corrosion rate however increases due to transpassive behavior on increasing oxidizer levels to very high values [2].

It is interesting to note that, once the passive film has been formed, it can be retained at oxidizer concentrations even lower than that needed for passive film formation.

Oxidizer concentration necessary to maintain passivity should at least be the same or higher than the required minimum to induce spontaneous passivation. There is also a concern of borderline passivity when any surface disturbance (scratching) will destabilize passivity, leading to increase in corrosion rate. The following conditions need to be kept in mind regarding passive behavior of metals and alloys [1, 2, 4]:

2 Principles of Corrosion Processes

- Corrosion rate is proportional to anodic current density in the active state irrespective of whether the metal or alloy exhibits passivity or not.
- Rate of cathodic reduction must exceed i_{crit} to achieve lower corrosion rates.
- Borderline passivity to be avoided while spontaneous passivation is preferred.
- Breakdown of passive films in high oxidizing environments or noble potentials due to transpassivity to be avoided.

2.19 Pitting Behavior of Passive Metals and Alloys

Chloride ions breakdown passivity or even at times prevent passivation of Fe, Cr, Ni, Co and stainless steels, since they can penetrate oxide films through pores and influence exchange current density. Breakdown of passivity by chloride ions is localized leading to pitting corrosion.

Susceptibility for pitting corrosion can be monitored through cyclic anodic polarization as shown in Fig. 2.30. Pitting potential can be experimentally determined. Initiation and propagation of pits occur between E_{pp} and E_{pit}. The potential where the loop closes during reverse scan is the protection potential (E_{prot}). New pits are initiated above E_{pit}. [1, 2, 4].

The following electrochemical aspects with respect to pitting initiation in active-passive metals and alloys may be noted.

- When E_{pit} and E_{prot} are the same, little pitting tendency
- If E_{prot} is more positive than E_{pit}, there will be no pitting.
- Repassivation tendency increases with high E_{prot}.

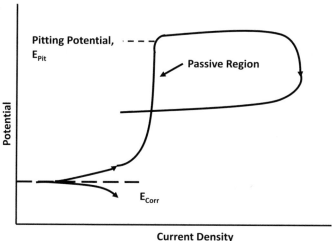

Fig. 2.30 Schematic cyclic anodic polarization of active-passive metal

- Pitting probability higher, if E_{prot} is more negative than E_{pit}.
- When E_{corr} is higher than E_{pit}, spontaneous pitting occurs
- E_{pit} lower, with higher concentration of aggressive anions like Cl⁻, which promote pitting.
- E_{pit} increases to nobler values with chromium and molybdenum addition when pitting corrosion is decreased.
- Pitting Resistance Equivalent Number (PREN) is used to measure relative pitting resistance of stainless steels in chloride media. [PREN = % Cr + 3.3% (Mo) + 16% (N)]

2.20 Anodic Protection [1, 2, 13]

Anodic protection through impressed anodic current can be applied to metals and alloys that exhibit active-passive behavior. The interface potential of the protected structure is increased to remain at the stable passive domain.

If an active-passive alloy such as stainless steel is maintained in the passive region through an applied anodic potential (or current), its initial corrosion rate (i_{corr}) can be shifted to a low value at i_p as shown in Fig. 2.31 [1, 2, 4].

Anodic protection unlike cathodic protection is ideally suited for protection of active-passive metals and alloys in aggressive environments such as high acidity and corrosive chemicals.

Typical anodic protection circuit is shown in Fig. 2.32 [2, 4, 13].

Anode is the container material itself (Eg: Stainless steel) Inert cathode materials having large surface area preferred. Recommended cathode materials for acid and

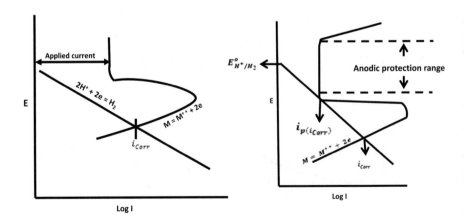

Fig. 2.31 Polarization curves illustrating anodic protection [2, 4]

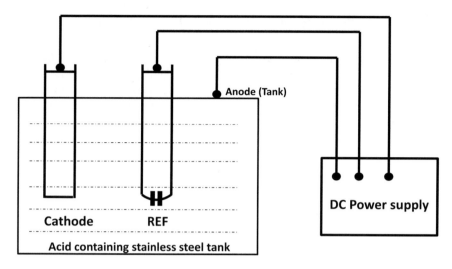

Fig. 2.32 Schematic illustration of anodic protection of acid containing stainless steel tank

corrosive industrial liquids include platinum-clad brass, chromium-nickel steel, silicon cast iron, copper, Hastelloy C and nickel-plated steel. Various types of reference electrodes such as Calomel, Ag/AgCl, Hg/HgSO$_4$ and platinum are used depending on the chemical environment.

Anodic protection can substantially reduce corrosion rate of active-passive alloys in very aggressive environments. For example, anodic protection of 304 stainless steels exposed to aerated sulfuric acid (5 M) containing about 0.1 M chlorides could reduce corrosion rate from an unprotected value of about 2000 μm/year, to about 5 μm/year. It has been widely applied to protect chemical storage tanks, reactors, heat exchangers and even transportation vessels.

A comparison between anodic and cathodic protection is given in Table 2.4.

Table 2.4 Comparison of anodic and cathodic protection methods [2–5]

Major factors	Anodic protection	Cathodic protection
Environmental conditions	For aggressive chemical corrosives	Moderate corrosion environments only
Suitability	Only for active-passive metals and alloys	Applicable to all metals in general
Operational features	Electrochemical estimation of appropriate protection range possible	Protective cathodic currents to be established through initial design and field trials
Cost aspects	Higher investment, but low operational costs	Low investment and higher operational costs

2.21 Cathodic Protection [1–5, 14, 15]

From potential—pH diagrams for metal—H_2O–O_2 equilibria, stability regions for metal immunity (state where the metal remains in its thermodynamically neutral form without oxidation) can be predicted. If the metal potential is maintained at its reversible equilibrium value where its remains in its neutral metallic state, corrosion can be eliminated.

Corrosion occurs due to differing potentials (anodic and cathodic areas) on a metal surface. When the potential differences are narrowed down and eventually eliminated with the entire surface converted to equipotential region (cathodic zone), corrosion can be arrested. Cathodic protection is based on the principle of corrosion control by making the metal surface cathodic (through cathodic polarization). This can be achieved either by attaching to a sacrificial anode or by an impressed DC current (potential). The entire corroding structure in forced to collect current (electrons) from the environment. Cathodic protection is defined as elimination or reduction of corrosion by making the entire metal a cathode by means of an impressed current or attachment to a sacrificial (more anodic) electrode. Basic electrochemical concepts involved in cathodic protection are illustrated in Fig. 2.33.

If the cathode ($E_{cathode}$) is polarized by an external current, the anodic reaction would be retarded, while the catholic reduction is enhanced. At the region, where the rates of cathodic and anodic reactions are equal, $E_{corr,}$ and I_{corr} are indicated defining the corrosion rate of the metal. If the corrosion potential is moved from E_{Corr} (B) to a lower value E through an applied current (C-D), rate of corrosion would decrease to (D-E) from I_{Corr}. If the applied current is further increased, corrosion current further deceases, as the potential moves to more active values. At

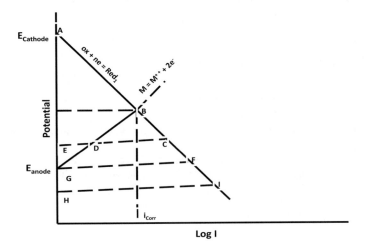

Fig. 2.33 Electrochemical illustration of cathodic protection [5]

an applied current equal to G-F, corrosion is nullified completely as the potential coincides with E_{anode}. Still higher applied currents (H-I) do not help (over-protection).

In natural aqueous environments at neutral pH as in sea water, reduction of oxygen ($O_2 + 2H_2O + 4e = 4OH^-$) is the cathodic reaction which is diffusion controlled. Polarization diagram representing cathodic protection under the above conditions is illustrated in Fig. 2.34. Corrosion rate under the conditions is dictated by limiting (diffusion) current for oxygen reduction and therefore, the applied cathodic current for corrosion protection is substantially lower.

The cathodic current density required to maintain the correct protection potential will vary depending on environmental conditions. The protection potential which is equal to the equilibrium potential for different metals can be calculated. For example, for iron,

$$Fe = Fe^{++} + 2e, E^0 = -0.44V \qquad (2.27)$$

Depending on the pH and from a knowledge of the solubility product (K_{sp}) of the reaction product ($Fe(OH)_2$ for example), concentration of Fe^{++} can be estimated and the corresponding equilibrium potential calculated. For neutral pH, the protective potential for iron works out to be about -0.62 V (NHE) or -0.85 to 0.90 V (Cu–CuSO$_4$).

Major applications of cathodic protection include external surfaces of pipelines, ship hulls, storage tank, jetties and harbor structures, steel sheets and piles, off shore platforms, floating subsea structures, reinforced concrete structures as well as water storage and circulating systems. Generally, exterior of pipelines and other structures

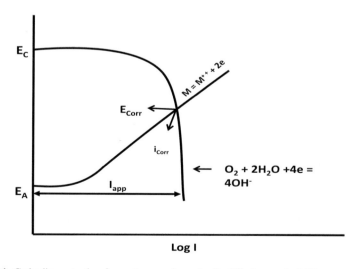

Fig. 2.34 Cathodic protection for systems under cathodic diffusion control [2]

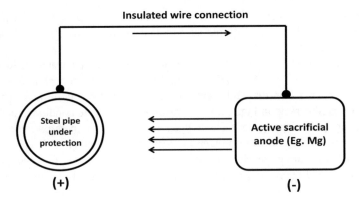

Fig. 2.35 Sacrificial anode cathodic protection

are initially well coated and cathodic protection is applied only to coating defects/holidays (5–10% of the damaged coatings).

Two methods of cathodic protection are:

(a) Sacrificial anode protection where the structure to be protected is connected directly to a more active metal/alloy (Fig. 2.35).
(b) Impressed current method, where the protected structure is connected to an auxiliary anode through a DC power supply (Fig. 2.36).

In sacrificial anode protection no external power source is used. More active anodes such as magnesium, aluminum and zinc can be used to protect steel structures which require only lower currents for protection in soils having low resistivity.

The following factors need to be ascertained in the choice of an appropriate galvanic anode.

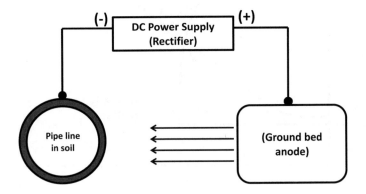

Fig. 2.36 Impressed current cathodic protection'

(a) Electrical energy content or anode capacity which is a measure of electric energy per weight provided by the sacrificial anode as dictated by Faraday's laws. Electrical energy content (EEC) for zinc anode is estimated to be 820 A hours/Kg which means that if a zinc anode were to discharge one ampere continuously, one Kg will be consumed in 820 h. The EEC output will be proportional to the anode efficiency and practical energy output will be actually dictated by the anode efficiency of the chosen galvanic anode.
(b) Solution potential
(c) Driving potential which is the difference between solution potential and the potential of the polarized structure.
(d) Anode bed—number of anodes in the back-fill.

The following parameters need to be predetermined for the application of impressed current protection.

(a) Source of DC Current.
(b) Estimation of current necessary for complete protection.
(c) Auxiliary anodes—choice, size, number, installation.
(d) Installation, design, erection and maintenance.

For large structures such as underground pipe lines, impressed current cathodic protection is used, while for smaller structures such as house-hold water tanks, ship's hull etc., sacrificial anodes can be effectively used. Initial coating of steel pipe lines and tubes can significantly reduce protection current requirements and thus save cost.

Approximate current requirements for cathodic protection of steel pipes are indicated below:

Uncoated structures in flowing sea water	10–15 mA/ft^2
Well-coated pipes in water	0.01–0.003 mA/ft^2
Excellently coated and exposed to water or under soil	0.0003 or less mA/ft^2

Some anode materials that could be used as ground-beds in impressed current cathodic protection are indicated in Table 2.5.

Design considerations for both impressed current and sacrificial anode systems have some common steps such as.

Table 2.5 Anode materials for impressed current cathodic protection

Anode material	Average consumption, kg/A-year
Cast iron	5–8
Steel scrap	5–10
Aluminum	4–6
Graphite	0.5–1.0
Lead	–
Platinum	–

(a) Exposed areas to be protected—exposed area at breaks and deteriorated coatings in case of coated structures.
(b) Polarized potential—Current density.
(c) Current demand—depend on the environment and nature of surface coating.
(d) Anode consumption—Numbers and weights of anode materials required to be determined from known consumption rates for the desired current demand. Anode number and distribution can be thus estimated. Anode resistance and design output current to be estimated.

Monitoring the effectiveness of cathodic protection of a structure is important. Potential measurements of a cathodically polarized structure with reference to a standard electrode can be periodically made. A potential of −0.85 V (Cu/CuSO$_4$) is sufficient for protection of steel in soil and natural water environments. The above criterion is not optimum and situations may arise when more negative potentials (up to −1.0 V) may be required for protection. Interference from IR components needs to be taken into account. Close Interval Potential Survey (CIPS) is an intensive monitoring technique. Direct Current Voltage Gradient (DCVG) method enables evaluation of protection and defects in insulation. Corrosion coupons (probes) can be used for monitoring of cathodic protection.

Controlled potential cathodic protection can be adapted to suit specific environments. For example, in sea-going vessels, the hull is subjected to variations in flow velocities which can alter limiting current density (for oxygen reduction), influencing cathodic protection current requirements from time to time. In such cases, controlling the potential (rather than current) would be more beneficial. **Controlled potential protection** can be used incorporating auxiliary anode—reference electrode attachment along with automatically—controlled power supply unit [2].

2.22 Stray Current Corrosion [1–5]

Stray-current corrosion is caused by several impressed current cathodic protection systems. In industrial protected systems, such as oil production industries having innumerable buried pipe lines, current leakage from impressed current anodes installed with cathodic protection systems can unintentionally enter a near-by unprotected structure (such as water pipelines) and leave from the surfaces creating severe corrosion (see Fig. 2.37).

Other sources of stray currents include DC electric power rail traction, welding units, electroplating cells and ground electric DC power.

Leaking stray currents from the above installations take a low resistance pathway to enter nearby unprotected structures before returning to the source. Regions from where current leaves are susceptible to stray-current corrosion, while areas receiving currents are protected!.

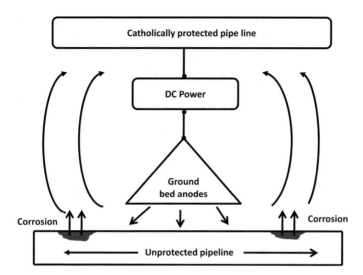

Fig. 2.37 Illustration of stray current corrosion [1]

Solutions to stray current corrosion include electrical bonding of the near-by unprotected structure. Simultaneously, additional anodes and increasing DC power capacity can accord full protection to all structures in the vicinity. Properly insulated couplings can also help reduce the problem (see Fig. 2.38). When impressed current protection systems are installed, anode ground beds should be so located to ensure that stray currents cannot leak to enter into other near-by structures.

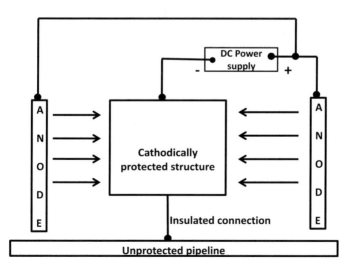

Fig. 2.38 Prevention of stray current corrosion by proper design

Direct stray currents can be classified as anodic, cathodic or a combined interference. Anodic interference occurs in close proximity to a buried anode. On the other hand, cathodic interference is encountered in close proximity to a polarized cathode, the potential shifting in a positive direction where current leaves the structure (causing corrosion damage). In combined interference, stray current pick-up occurs close to anode and discharge closer to cathodically polarized areas. The corrosion damage could be severe in this case since current pickup (overprotection) and discharge (corrosion) are both detrimental.

2.23 Biofouling and Microbially Influenced Corrosion [1, 16–22]

Biofouling and microbially influenced corrosion (MIC) brought about by various microorganisms has assumed great industrial significance in recent years. While biofouling involving attachment of microorganisms onto metals and alloys promotes MIC, electrochemical principles govern biological catalysis of oxidation/reduction reactions. The role of microorganisms in metallic corrosion can be seen as a biological catalyst influencing anodic or cathodic reactions along with generation of biogenic corrodants.

Various microorganisms as listed below are implicated in various MIC processes.

Sulfur/sulfide oxidizing bacteria	*Acidithiobacillus thiooxidans*
	Acidithiobacillus ferrooxidans
Sulfate reducing bacteria (SRB)	*Desulfovibrio* spp.
	Desulfobacter spp.
	Desulfotomaculum spp.
Iron/manganese oxidizing bacteria	*Gallionella*
	Crenothrix
	Leptothrix
Metal-reducing bacteria } Slime producing bacteria }	*Pseuclomonas* spp.
	Bacillus spp.

Fungi such as *Cladosporium resinae, Aspergillus niger*

Microorganisms can promote corrosion through changes in electrochemical conditions at the metal-solution interface which is modified and conditioned by prior biofilm formation. Microbial adhesion to metal surfaces as well as interaction with extracellular polymeric substances (EPS) promote MIC.

The sulfur-bacteria cycle in nature is closely linked to MIC. Sulfur- Sulfide-iron oxidizing aerobic *Acidithiobacillus* bacteria bring about oxidation of sulfur, ferrous ions and various sulfides to ferric-ion containing sulfuric acid, while anaerobic sulfate-reducing bacteria such as *Desulfovibrio* sp. reduce sulfates to sulfides promoting formation of H_2S and metallic sulfides. Fungi such as *Aspergillus niger* and *Cladosporium resinae* produce organic acids such as citric acid which can dissolve many ferrous and non ferrous metals and alloys. Potential-pH diagrams can be used to demarcate stability and activity regions of various microorganisms as well as to understand metal corrosion behavior induced by bacterial activity. While anodic reaction is oxidation of the metal, the cathodic reactions could be reduction of hydrogen or oxygen depending on pH of the environment.

Corrosion-promoting microbiological functions are listed below:

- Production of organic and inorganic acids, sulfides, phosphides and ammonia.
- Biodegradation of surface coatings, passive films and inhibitors.
- Formation of oxygen concentrations cells due to heterogeneous biofilm formation, generating localized anodes and cathodes promoting pitting and crevice corrosion.
- Direct participation in anodic and/or cathodic electrochemical reactions, influencing corrosion kinetics.

Microorganisms thus do have both direct and indirect roles in causing metallic corrosion.

MIC of various steels brought about by Sulfate Reducing Bacteria (SRB) such as *Desulfovibrio sulfuricans, Desulfovibrio vulgaris* and *Desulfotomaclum nigrificans* has been extensively studied. SRB are the most widely implicated anaerobe causing corrosion in oil and gas, nuclear power, water treatment, mining and chemical-based industries. Pipe line internal corrosion and corrosion of sea-going vessels and marine installations are caused by the presence of these anaerobes.

There are several proposed mechanisms illustrating the role of SRB in steel corrosion

The cathodic depolarization hypothesis proposed in 1934 [23] dealt with bacterially-induced depolarization of hydrogen reduction reaction by hydrogenase-positive SRB as per the following reaction pathways.

$$Fe = Fe^{++} + 2e \text{(corrosion of iron)} \quad (2.27)$$

$$H_2O = H^+ + OH^- \quad (2.82)$$

$$H^+ + e = H_{ads} \quad (2.83)$$

$$2H_{ads} = H_2 = 2H^+ + 2e \text{ (cathodic reaction)} \quad (2.84)$$

$$SO_4^{--} + 8H^+ + 8e = S^{--} + 4H_2O \text{ (Bacterial sulfate reduction)} \quad (2.85)$$

$$Fe^{++} + S^{--} = FeS \text{ (Biogenic reaction product)} \quad (2.86)$$

Iron hydroxide and iron sulfide precipitates coat the corroded steel surfaces.

The above hypothesis has since been questioned and many alternative mechanisms proposed. The role FeS formed during MIC has been highlighted as a potential promoter of corrosion through galvanic interaction (Fe–FeS couple), cathodic depolarization and stress initiation.

Corrosion fatigue and hydrogen embrittlement of steel could be caused by SRB activity. H_2S—produced by the bacteria can decrease pH at the metal-biofilm interface (causing acidification) and promote hydrogen permeation into the metal. Even anodic depolarization by iron-reducing and sulfate-reducing bacteria can be caused in the MIC of steels Electrical Microbially influenced Corrosion (EMIC) was proposed to substantiate metallic iron as an electron donor, while biogenic H_2S aggravates the corrosive effect [19]. Anodic iron oxidation can be enhanced by electron consumption through bacterial sulfate reduction. Another deleterious role of marine and soil sulfate reducing bacteria is their ability to destabilize passive films present on active-passive alloys such as stainless steels. Corrosion behavior of stainless steels is significantly influenced by biofilms. Ennoblement of stainless steels has been reported in the presence of biofilms containing iron-and manganese-oxidizing bacteria [1, 16–22].

Tubercle formation in oil and water pipelines is a serious problem, mainly brought about by iron-oxidizing bacteria. Massive tubercles can seriously impede mass and heat transfer in pipes and heat exchangers. In various fuel systems, microbial growth can create problems such as clogging of valves, filters and pipes, sludge accumulation, corrosion of storage tanks, biodegradation of hydrocarbon oils and breakage of engine parts.

Microbial corrosion in aircraft components has now been well established. Various bacterial and fungal species can grow and proliferate in fuel tanks. Bacteria like *Pseudomonas* and fungi such as *Cladosporium resinae* have been isolated from aircraft aluminum alloy fuel tanks.

Intergranular cracking and extensive pitting of aluminum alloys can occur due to microbial activity.

Biofouling and microbial corrosion are extensively observed in reinforced concrete structures and in human body implants. MIC of reinforced steels in concrete is a serious problem since integrity of bridges, buildings, terrestrial and marine environment-structures could be compromised. Formation of biogenic corrosion products inside concrete components induce internal stresses causing catastrophic fracture.

Human body environment has been proved to be corrosive, especially with respect to various implant biomaterials such as stainless steels and titanium alloys. An oxygenated saline electrolyte at neutral pH containing water, chlorides, calcium, phosphates, amino acids and various organic polymers present a corrosive environment promoting anodic and cathodic electrochemical reactions. Various bacterial species inhabiting human body organs and circulatory systems interact with implanted metals and alloys. Corrosion forms such as pitting, crevice corrosion, galvanic attack, stress corrosion and intergranular attack are known to be prevalent in different locations of the human body. Some examples are given below:

(a) Austenitic stainless steels used in cranial plates, orthopaedic and spinal implants undergo pitting and crevice corrosion. Stress corrosion and corrosion fatigue has been observed in orthopedic joint implants.
(b) Cobalt-chromium alloys used in dental implants, orthopaedic fracture plates and joint replacements undergo dissolution in presence of serum proteins causing metal toxicity.
(c) Titanium alloys used as cranial and orthopaedic fracture plates undergo fracture.

2.24 Summary

Various corrosion processes can be explained with respect to electrochemical and microbiological mechanisms. Thermodynamic and kinetic aspects of corrosion are explained with reference to galvanic and concentration cells, electrochemical polarization, electrode kinetics and mixed potential theory. Mass and electron transport processes at a metal-solution interface can be understood in terms of concentration and activation over-potentials and kinetic equations derived to estimate corrosion rates. Cathodic protection, the extensively used protection method is based on principles of immunity and cathodic polarization. Principles of passivity find application in anodic protection of active-passive metals and alloys exposed to aggressive acid environments. Corrosion processes are catalyzed by various soil and marine microorganisms. Electrochemical mechanisms control microbially influenced corrosion, promoted by formation of biofilms. Biomaterials used in human body implants are exposed to corrosive environments prevailing in the human body.

Most of the types of corrosion occurring in industrial environments can be effectively explained in terms of electrochemical and microbiological principles.

Acknowledgements The author is thankful to The National Academy of Sciences, India (NASI) for the award of Honorary Scientist Contingency Grant.

References

1. Natarajan KA (2012) Advances in corrosion engineering, national program on technology enhanced learning (NPTEL). https://nptel.ac.in
2. Jones DA (1996) Principles and prevention of corrosion. Prentice Hall, NJ
3. Revie RW (2011) Uhlig's corrosion handbook, 3rd edn. Wiley, Hoboken
4. Fontana MG (2005) Corrosion engineering. Tata McGraw-Hill, New Delhi
5. Revie RW, Uhlig HH (2008) Corrosion and corrosion control. Wiley Interscience, Hoboken
6. Zhang XG (2011) Galvanic corrosion. In: Revie RW (ed) Uhlig's corrosion handbook, 3rd edn. Wiley, Hoboken, pp 123–143
7. Bimetallic corrosion. From web www.npl.co.uk/upload/pdf/bimetallic2007105114556.pdf

8. https://www.researchgate.net/figure/Anodic-index-of-metals-2th/1288427213
9. Electrochemical Kinetics of Corrosion and Passivity. https://www.staff.tugraz.at/robert.schennach/Electrochemical_Kinetics_of_Corrosion_and_passivity.pdf
10. Mixed potential Theory. www.Uobabylon.edu.ig/eprints/publication_12_18241_228.pdf
11. Staehle RW (1968) Fundamental aspects of corrosion of metals in aqueous environments. Special Lecture Series. University of Minnesota, Minneapolis
12. Greene ND (1962) Predicting behavior of corrosion resistant alloys by potentiostatic polarization methods. Corrosion 18:136t–142t
13. Edeleanu C (1960) Corrosion control by anodic protection. Platin Met Rev 4:86–91
14. Francis PE. www.resource.nple.co.uk/docs/science_technology/materials
15. Ashworth V. Principles of cathodic protection. https://booksite.elsevier.com/brochures/shreir/pdf/principles_of_cathodic_protection
16. Natarajan KA (2018) Biotechnology of metals-principles, Recovery Methods and Environmental Concerns. Elsevier, Cambridge
17. Borenstein SW (1994) Microbiologically influenced corrosion handbook. Woodhead Publishing India Pvt. Ltd., Cambridge
18. Gaylarde CC, Videla HA (1995) Bioextraction and biodeterioration of metals. Cambridge University Press, Cambridge
19. Enning D, Garrelfs J (2014) Corrosion of iron by sulfate-reducing bacteria: new views of an old problem. Appl Environ Microbiol 80:1226–1236
20. Loto CA (2017) Microbiological corrosion: mechanism, control and impact—a review. Intl J Adv Manuf Technol 92:4241–4252
21. Beech IB, Gaylarde CC (1999) Recent advances in the study of biocorrosion—an overview. Rev Microbiol 30:177–190
22. Little BJ, Lee JS (2014) Microbiologically influenced corrosion: an update. Intl Mater Rev 59:384–393
23. Kuhr VWCAH, van der Vlught LS (1934) The graphitisation of cast iron as an electrochemical process in anaerobic soil. Water 18:147–165

Chapter 3
Corrosion Sensing

Jeff Demo and Ravi Rajamani

Abstract Sensing is a key part of diagnostics and prognostics of corrosion. In this chapter we will outline the basic techniques used to detect corrosion; without going into details of the analysis that would accompany these techniques. That is covered elsewhere in this book; see Chap. 5 on Data Analytics for Corrosion Assessment. Corrosion is an electrochemical reaction of the metal to the environment, so it can be assessed either by directly detecting it on the material or by estimating its likelihood of occurrence indirectly via measuring ambient conditions. Both will be covered in this chapter. Lesser known phenomena involving corrosion from microorganisms are dealt with in Chap. 2 on the Principles of Corrosion Process.

Keywords Corrosion sensing · Environmental monitoring · Electrochemical · Free corrosion · Galvanic corrosion · Inspection · LPR · Radiography · Strain measurement · Optical corrosion sensing

3.1 Introduction

Corrosion is the result of a metal's reaction to environmental conditions and producing structurally weaker oxides. It starts off as a local phenomenon on the surface and spreads if unchecked. But it can also manifest itself internally along imperfections such as grain boundaries, or interfaces between metal layers (see Chap. 2). Several broad changes take place that can be detected by appropriate sensors and correlated to the presence of corrosion. Because this is an electromechanical phenomenon, several techniques rely on the detection of the electric currents that are generated by the reactions. Others rely on detecting chemical changes that indicate the presence of the products of oxidation. And yet others directly detect corrosion

---

J. Demo (✉)
Luna Innovations, Roanoke, VA, USA
e-mail: demoj@lunainc.com

R. Rajamani
drR2 Consulting, West Hartford, Connecticut, USA

© Springer Nature Switzerland AG 2020
G. Vachtsevanos et al. (eds.), *Corrosion Processes*, Structural Integrity 13,
https://doi.org/10.1007/978-3-030-32831-3_3

visually by observing the images of metal surfaces and apply pattern recognition techniques, or profiling the surface for imperfections engendered by corrosion. With the growth of computational power, these techniques can be quite reliable. Chapter 5 discusses the analytical techniques used to detect and classify corrosion from images.

In the chapter we will categorize sensing broadly in terms of the changes to physical parameters that corrosion affects. These modalities are electrochemical, electrical, optical and visual, mechanical, chemical, and radiation.

Electromechanical methods are most direct because corrosion is fundamentally an electrochemical process that acts on the micro scale where the metal comes in contact with an oxidizing agent, which is typically the surrounding electrolytic medium. Many techniques have been developed to measure the electrical properties of the medium as well as the actual material to assess whether the chemical reactions associated with corrosion are taking place, and possibly the magnitude of these reactions.

Corrosion effects a physical change in dimensions of the metal, both in the thickness and in the surface finish. These mechanical characteristics can form another means of assessing corrosion's initiation and growth. These can be actual dimensional measurements as well as visual measurements.

In addition to direct measurements, indirect means provide a useful way to estimate the presence and progression of corrosion and corrosive conditions. For example, using a sacrificial coupon of the same material as the structure, exposed to the same environment as the structure, is an oft-used method for corrosion assessment. Another powerful way is to analytically assess the propensity of the environment to initiate and encourage corrosion by analyzing its physical and chemical properties and employing a model-based method for assessing corrosion. In the following sections we will analyze each of these methods in more detail. Some of these methods require considerably more analysis, and those details can be found elsewhere in this work, such as in Chap. 5.

3.2 Electrochemical/Electrical Techniques

At a base level, corrosion is an electrochemical process, and as such, corrosion monitoring and evaluation can be performed through a number of electrochemical and electronic sensing techniques. These techniques can be used in both laboratory and field environments, though some methods are better suited for one over the other. Potential measurements, for example, are generally best suited for laboratory settings where laboratory equipment and a well-controlled environment are available. Potential measurements, including both potentiostatic and potentiodynamic techniques, provide insights into electrochemical conditions that result in metallic corrosion. As corrosion requires the transfer of electrons and ions between substrates and electrolytes, corrosion rates are equivalent to electrical currents. As such, the rates of these corrosion processes are strongly dependent upon potential differences between the metals and the surrounding electrolyte [1].

3.2.1 Potentiostatic and Potentiodynamic Evaluation

Potentiostatic evaluation of metals is generally a lab-based corrosion evaluation technique, often relying upon potentiostat equipment to accurately control the potential of a Counter Electrode (CE) against a Working Electrode (WE). This allows a well-defined measurement of the potential difference between the WE and a Reference Electrode (RE). Similarly, galvanostatic testing is performed by controlling the current flow between the WE and the CE. In both scenarios, the material of interested acts as the WE, all electrodes are exposed to an electrolyte, and corrosion potential (E_{corr}) or corrosion current (i_{corr}) is monitored.

Using a potentiostat, cathodic and anodic currents can be measured for the corrosion cell by applying and holding electrical potentials on the cell. Plotting these currents, along with the measured cell current at each potentiostatic step provides an overall picture of the electrochemical processes associated with corrosion under conditions present in the electrochemical test cell (Fig. 3.1). As shown in the image, the vertical axis represents applied electrical potential, and the horizontal axis represents the logarithm of absolute current for a generic material. Straight lines labeled as Cathodic and Anodic currents represent the theoretical currents for the material, and the curved line shows the total current, i.e. the sum of anodic and cathodic currents. The value of either anodic of cathodic current at their intersection is i_{corr}, the corrosion current, a direct measurement of corrosion rate of the material.

Potentiodynamic scans, wherein applied potentials are dynamically scanned across the range of interest, provide the same information as a sequence of potentiostatic measurements. The critical information collected from this testing are the E_{corr} and i_{corr} values that can be converted to corrosion rates for the material under test. Knowing the i_{corr} value, corrosion rate can be calculated using Faraday's law:

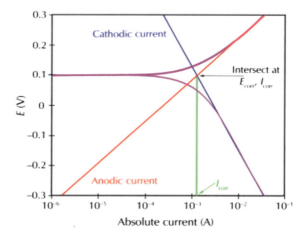

Fig. 3.1 Electrochemical interactions in a potentiostatic/potentiodynamic test cell (https://www.gamry.com/application-notes/corrosion-coatings/basics-of-electrochemical-corrosion-measurements/)

$$Q = nFM \tag{3.1}$$

where Q is the charge in coulombs resulting from the corrosion reaction, n is the number of electrons transferred per molecule or atom of the material under test, F is Faraday's constant (96,485 coulombs/mole) and M is the number of moles of the reacting material under test. An alternate configuration of Faraday's law incorporates the concept of equivalent weight (EW), which is defined as

$$EW = \frac{AW}{n} \tag{3.2}$$

where AW is the atomic weight of the species. As M = m/AW, where m = the mass of the test materials under reaction,

$$m = \frac{(EW)Q}{F}. \tag{3.3}$$

In terms of corrosion rate (CR), this equation can be rewritten as

$$CR = \frac{i_{corr} K * EW}{dA} \tag{3.4}$$

where i_{corr} is the measured corrosion current in amperes, K is a constant defining the units for the corrosion rate (i.e.3272 mm/[A-cm-year], 1.288 × 10^5 milli-inches/[A-cm-year]), EW is equivalent weight in grams/equivalent, d is material under test density in g/cm^3, and A is sample area in cm^2 [2].

3.2.2 Electrochemical Impedance Spectroscopy and Polarization Resistance

While potentiostatic and potentiodynamic scans are generally suitable for laboratory material evaluations where potentiostat devices are available, they do not lend themselves to field applications where more robust, lower power instrumentation may be required. The same concepts as defined for laboratory techniques can, however, be leveraged for field-level analysis of corrosion rates. Potentiostatic and potentiodynamic scans offer a full picture of anodic and cathodic currents across a wide range of electrochemical potentials, however, by focusing on a small potential range surrounding the E_{corr} value, assumptions can be made to assess corrosion potential in a means suitable for embedded corrosion monitoring systems.

Electrochemical impedance spectroscopy (EIS) uses an alternating current (AC) signal applied across a set of electrodes fabricated out of the material of interest to determine corrosion rate of the material under test at one or more excitation frequencies. By exciting the electrodes around the E_{corr} value, and using

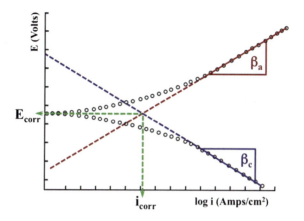

Fig. 3.2 Determination of material Tafel constants (https://www.gscsg.com/tafel-extrapolation.html)

known Tafel constants (defined in the discussion related to Fig. 3.2) for a given material of interest, corrosion rates can be calculated through restriction of potential application to only a small range of applied potentials. By applying a known excitation potential and measuring the resulting current, a polarization resistance (R_p) can be calculated as the slope of the current-versus-voltage curve, which approximates a straight line across the small excitation range around E_{corr}. Linear polarization resistance (LPR) sensors are a popular embodiment of these principles for corrosion detection, especially for liquid environments.

Coupled with the R_p values, specific material properties associated with the material of interest must be known or estimated a priori for determination of corrosion rates. Specifically, anodic and cathodic Tafel constants for the material of interest must be used. Recalling the potentiostatic and potentiodynamic scan techniques described, the Anodic Tafel constant (β_a) and the Cathodic Tafel constant (β_c) are defined as the slopes of anodic and cathodic currents that pass through the intersection of E_{corr} and i_{corr} during potentiodynamic scans (Fig. 3.2).

R_p and i_{corr} are related as defined in Eqs. 3.5 and 3.6, where B is the Stern-Geary constant and β_a and β_c are anodic and cathodic Tafel constants specific to the electrochemical processes for corrosion of the alloy of interest. Once i_{corr} is determined, Eq. 3.4 can be used to determine corrosion rate for a specific sensor configuration.

$$R_p = \frac{B}{i_{corr}}; \quad or \quad i_{corr} = \frac{B}{R_p} \quad (3.5)$$

$$B = \frac{\beta_a \beta_c}{2.3(\beta_a + \beta_c)} \quad (3.6)$$

As stated, specific electrical excitation requirements must be adhered to for accurate corrosion current measurements using EIS and polarization resistance techniques. Small variations of the applied potential around E_{corr} can be made with sine wave excitations, triangle wave excitations, or potential step excitations.

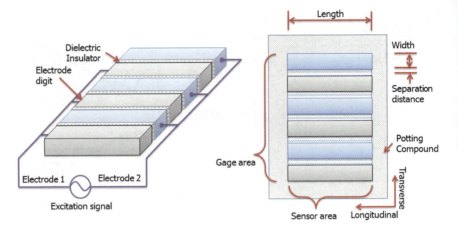

Fig. 3.3 Schematic isometric (left) and top (right) views of a laminated interdigitated two-electrode sensor. The two electrodes are formed by electrically connecting every other digit to form the interdigitated electrodes. Electrodes can be fabricated with a single materials for free corrosion measurements or different materials for galvanic corrosion measurements

For sine wave excitation, voltages of not more than 30 mV amplitude around E_{corr} applied at a frequency between 0.01 and 10 Hz are suitable. It is advisable to average multiple sine wave cycles to increase confidence in the measurement. Triangle wave excitations may be used as well, again at amplitudes not more than 30 mV around the E_{corr} value. Ramp rates of 0.045 mV/s to 10 mV/s are suitable for triangle wave excitations. Potential step excitations may also be used, wherein discrete potential steps both above and below E_{corr} are used. These steps should remain within ±30 mV of E_{corr} to obtain a linear fit of the current response. At each step, the hold time at a given potential should be sufficient to obtain a steady state measurement, ideally using the average of multiple measurements at each potential hold [2, 3]. These excitations can be applied to a two-electrode sensor configuration as shown in Fig. 3.3.

3.2.3 Galvanic Corrosion

Electrochemical techniques described above work well for determination of free material corrosion, wherein a single alloy of interest is undergoing corrosion processes. However, in practice, there are often multiple alloys in single electrochemical cells, resulting in galvanic corrosion processes. In these cases, two dissimilar metals, one acting as an anode and the other acting as a cathode, form a galvanic couple which drives corrosion. As with free corrosion processes, the current flow between the two dissimilar metals defines the overall corrosion rate of the cell. Galvanic corrosion sensors can be configured in the same manner as free corrosion sensors, with the difference of having alternating sensor digits fabricated out of these dissimilar materials (Fig. 3.3). Corrosion current of the electrochemical cell can be made using a number of methods, though the most straightforward methods consist of the use of a zero resistance ammeter (ZRA) or a precision resistor. Once galvanic currents are measured, current measurements can be converted to mass loss via Faraday's law.

A ZRA circuit will apply a current bias to control the potential difference between the two dissimilar metal electrodes in a galvanic couple. As a galvanic corrosion sensor may consist of two electrodes of dissimilar metals, the ZRA must connect to each electrode and provide a positive current to the more negatively charged electrode. The current required for the potential difference of the electrochemical cell to reach zero is the galvanic corrosion current of the electrochemical cell.

An alternative to using a ZRA to measure galvanic corrosion current is to place a precision resistor between the two electrodes of dissimilar metals. Using Ohm's law (V = IR) with a known resistance (R) and a measured voltage (V) the galvanic current (I) can be measured. It is recommended that the value of the precision resistor used in this circuit be selected to be as low a value as possible while still allowing for a high precision voltage measurement to be obtained [4].

3.2.4 Electrochemical Noise

Electrochemical noise techniques are based on the current and potential fluctuations that occur during corrosion processes. Random electrochemical processes occurring on a corroding substrate will generate noise in the overall potential and current signals. Specific types of corrosion, including general corrosion, pitting, crevice, and stress corrosion cracking, will each have a characteristic noise signature that can be monitored in situ. This signature can then be used to evaluate the types of corrosion occurring on a substrate and the severity of the underlying corrosion processes [5]. Electrochemical noise is the result of stochastic pulses of current generated by sudden film rupture, crack propagation, discrete events involving metal dissolution and hydrogen discharge with gas bubble formation and detachment. The technique of measuring electrochemical noise uses no applied external signal for the collection of experimental data.

As with potentiostatic and potentiodynamic test configurations, electrical noise measurement setups generally require a set of three electrodes, with a mix of reference, counter, and working electrodes (RE, CE, and WE, respectively). Data acquisition systems for electrical noise evaluations must be configured for micro-volts (μV) or lower level resolutions, as the signals of interest tend to be relatively small. Data acquisition systems should also allow for simultaneous measurements of voltages and currents within the three electrode system.

Analysis of electrochemical noise signals are generally performed as a statistical analysis of both voltage (potential) and current measurements within a system.

Table 3.1 Electrochemical noise measurement analysis to determine corrosion mechanisms [5]

Mechanism	Potential		Current	
	Skewness	Kurtosis	Skewness	Kurtosis
General	<±1	<3	<±1	<3
Pitting	<−2	≫3	>±2	≫3
Transgranular SCC	+4	20	−4	20
Intergranular SCC #1	−6.6	18–114	1.5–3.2	6.4–15.6
Intergranular SCC #2	−2 to −6	5–45	3–6	10–60

Statistical parameters such as the mean, variance, third and fourth moments (i.e., skewness, and kurtosis) of measurements are often used to evaluate collected electrical noise data. For example, Reid and Eden [6] offer an analysis of potential and current skewness and kurtosis to determine the corrosion mechanisms based on electrochemical noise (Table 3.1).

It is also possible to couple electrical noise measurements to other electrochemical monitoring techniques such as polarization resistance measurements. It has been shown that the ratio of electrochemical current noise to general corrosion current (i_{corr}) can be used as an improved indication of localized corrosion [5, 7].

3.2.5 Electrical Resistance

Electrical resistance (ER) corrosion sensors operate on the principle of measuring the change in electrical resistance of a thin layer of metal or alloy over time as a result of corrosion. When exposed to a corrosive environment, corrosion related material loss will thin the alloy layer, resulting in an increase in electrical resistance. As the resistivity of an alloy is dependent upon temperature, electrical resistance sensors require a protected "reference" sensor exposed to the same thermal environment, but physically shielded from the surrounding corrosive environment. The reference sensor will remain pristine while the exposed element will be impacted by corrosivity of the surrounding environment. By quantifying the change in resistance relative to an initial pristine state or some prior reading, cumulative corrosion measurements can be made, often using a sensor manufacturer-defined correlation between changes in impedance and alloy mass loss. While free corrosion sensors and galvanic corrosion sensors offer instantaneous corrosion rate measurements, electrical resistance sensors provide cumulative corrosion measurements for a defined time period. The average corrosion rate for a given time period can be estimated by dividing cumulative corrosion by the time interval [8]. This is an example of a sacrificial coupon sensor, more details are discussed below. Another example of the use of this principle is described next with the inductive shift sensor.

3.2.6 Inductive Shift

Similar to electrical resistance probes, inductive shift sensors allow for identification of corrosion by monitoring the change in inductance of a coil coupled to a sacrificial component. As with electrical resistance measurements, inductive shift sensors are susceptible to environmental impacts such as temperature, and will benefit from the use of a protected reference sensor for compensation. The reference sensor must be physically isolated from corrosive conditions while remaining in the same thermal environment as the sacrificial sensor. Inductive shift sensors tend to be more sensitive to the initiation of corrosion processes than their electrical resistance counterparts with similar total corrosion spans.

3.3 Environmental Sensing

The electrical and electrochemical sensing techniques discussed so far all either measure actual material loss or electrical processes directly associated with the processes of corrosion. However, it is also possible to monitor environmental effects to make assumptions about the corrosivity of an environment, and thus, the likelihood of corrosion to occur.

3.3.1 Relative Humidity and Temperature

Relative humidity sensors provide a measurement of the amount of water vapor in air expressed as a percentage of the amount needed for saturation at the same temperature. Relative humidity is a key driver of corrosion as it is a measure of the availability of moisture that can generate thin films of electrolyte on material surfaces. Further impacting the formation of electrolyte layers on material surfaces is temperature of both the ambient environment and the surface of interest.

Direct measurements of temperature are used to characterize the local environment and can be used in combination with other parameters to determine if corrosion is expected. Temperature measurements may also be used for compensating other sensing elements and interface electronics which can have temperature dependence.

When combined with surface or air temperature readings, relative humidity measurements provide dew point temperature. Dew point is a valuable parameter due to the fact that it can indicate when a substrate is in a condensing environment under which corrosion processes will occur.

Relative humidity (RH) data is used to calculate the ISO time of wetness by summing the time RH is greater than 80% and temperature is above 0 °C, and dividing by the total exposure time. Classification of the time of wetness category is

Table 3.2 Environmental classification for percent time of wetness according to ISO 9223

Category	Time of wetness (%)
τ_1	$\tau \leq 0.1$
τ_2	$0.1 < \tau \leq 3$
τ_3	$3 < \tau \leq 30$
τ_4	$30 < \tau \leq 60$
τ_5	$60 < \tau$

defined by percent time of wetness over the full exposure period (Table 3.2). Environmental severity is classified as "very low" under category τ_1 though very high under category τ_5.

Relative humidity and temperature measurements can be made using a wide array of commercially available sensors based on resistive, capacitive or alternate methods of parameter detection [9, 10].

3.3.2 Environmental Contaminants

Evaluation of atmospheric contaminants can provide valuable information relating to alloy corrosivity. Corrosion and environmental severity measurement systems may include one or more of the following sensor modalities to supplement an understanding of the local microclimate, to characterize environmental exposure throughout a structure's lifetime, and to identify specific geographical locations in which accelerated corrosion may occur.

- Solution Conductance—Solution conductance (or surface conductance) measurements can be used to measure moisture and contaminants, such as marine salts or aerosols, that form electrolyte layers on the surface of a structure. The hygroscopic properties of the contaminant deposits such as salts will impact the moisture layer and therefore conductance measurements. Conductive measurements are indicative of conditions that may promote corrosion and are obtained using two-electrode sensors. Sensors consist of two interdigitated electrodes fabricated of a single noble alloy such as gold, separated by an electrical insulator (Fig. 3.4). The separation distance between electrodes should be uniform and should not exceed 300 μm. Electrical measurements are taken with the sensor by measuring current response of the sensor to a known applied sinusoidal excitation signal. The applied excitation signal should be a small signal AC voltage, nominally around 30 mV in amplitude, and corresponding measured currents will be dependent upon the excitation voltage selected as well as the surrounding environment. Excitation voltages should be applied at frequencies between 0.1 and 1000 kHz. Conductance sensors can also be used to evaluate the time of wetness (TOW) of a surface based on a wet/dry threshold

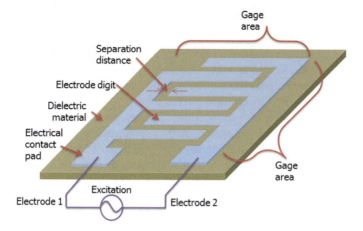

Fig. 3.4 Schematic of a thin film interdigitated two-electrode sensor. The electrodes can be fabricated using a noble material such as gold for use in a solution conductance sensor element

for conductance, above which a surface is considered "wet" and below which a surface is considered "dry." TOW can be used in a number of analysis and modeling techniques for corrosion evaluations.
- Sulfur Dioxide (SO_2)—Sulphur dioxide dissolves readily in water, resulting in a weak, corrosive acid. The effects of atmospheric SO_2 exposure far exceed the effects of other gaseous atmospheric contaminants and should be considered a priority over other gas contaminant measurements.
- Nitrogen Dioxide (NO_2)—As with SO_2, atmospheric NO_2 combines with water vapor to create a weak nitric acid. As NO_2 is a byproduct of many industrial processes and a common pollutant, characterization of this contaminant can provide insight into exposure resulting from environmental exposure and operations in close proximity to a large contingency of industrial processes.
- Ozone (O_3)—O_3 acts as a strong oxidizer that can lead to accelerated material degradation at increased exposure levels. As ozone levels increase with altitude, certain structures including aircraft, can be exposed to higher levels of ozone than other ground-based structures and equipment.
- Hydrocarbons—Exposure to fluid leaks such as JP-8 fuel or hydraulic fluids can degrade protective coatings and seals and drive corrosion. Identification of aromatic hydrocarbons in a microclimate can provide indications of loss of coating protective properties and corrosive conditions.

The above selection of chemical contaminant sensors is representative of the types of additional sensors that can be included in a comprehensive corrosion monitoring system. All sensor types are not required for the implementation of such a system, but may add value based on the additional information they provide regarding corrosivity within a structure [11].

3.4 Mechanical Methods

These methods rely on measuring dimensional parameters and assessing the extent of corrosion by the sensed changes; methods borrowed from the rich history of non-destructive evaluation (NDE). The result of corrosion on external surfaces is the oxidation of material by uniform corrosion across a large portion of the surface area, or localized corrosion in spots or pits. Internal corrosion can occur in certain metals buried in other material such as rebar in concrete, or at a microscopic level within the metal, along grain boundaries, or at interfaces of certain dissimilar metals or alloys. When certain dissimilar metals are joined, there is a higher tendency for galvanic action to encourage corrosion at the boundary. When buried, the oxidation of metal will not result in a loss, but instead in the change of volume of the structure, or the increase in forces acting on the surrounding matrix. Measuring these changes can be a proxy for measuring corrosion. Thickness and mass are most readily measured parameters using any number of probes. Other methods such as the direct observation of defects can be used as well to assess corrosion. Surface defect measurements can be done in many ways: Ultrasonic, x-ray, thermography, and white light interferometry.

3.4.1 Ultrasonic Probes

This is a common way of measuring thickness in metals. The accuracy, resolution, and depth of measurements are governed by the strength and the frequency of the acoustic signal as well as the coupling between the sensor and the metal, which is typically oil or water.

Typically, the frequency of the ultrasound signal ranges from 500 to 25 MHz, with the higher frequency signals able to detect smaller defects. These probes are straightforward to describe, if not easy to employ everywhere. The sensing system consists of a source of ultrasonic waves, typically made with piezoelectric elements, excited by an AC signal generally shaped to form a pulse. The return signal is sensed by the same probe (acting in a pulse-echo mode) or by a separate probe (acting in a pitch-catch mode). The sensor head is held against the structure and acoustically coupled to the surface with an appropriate coupling medium such as water or oil. This is needed to ensure that the energy directed inside the metal is not dissipated in air. Knowing the speed of sound inside the material and the time of flight of the pulse makes it easy in principle to calculate the thickness. There are frequency-based techniques as well, that depend on how easily continuous signals of different frequencies propagate through metals of different thicknesses, but that is not as common. In fact, the ISO standard on ultrasonic testing (UT) for thickness measurements only lists the pulsed methods (both relying on wave reflection and wave transmission [12].

Fig. 3.5 Phased array ultrasonic inspection of pipes (GE inspection technologies)

While measurement systems that are portable are most common because of cost and convenience (Fig. 3.5), permanently installed devices can provide more precise long-term measurements that allow the trending of dimensional parameters over time [13]. Of course, the disadvantage of a fixed probe is that the area being measured is limited in extent [14]. UT systems can also be configured to have the source and the sink on opposing surfaces of the structure, but that makes the system more cumbersome and not particularly suited for corrosion detection. Significant disadvantages of UT are the time and effort it takes, and the need for experienced operators.

While these are standard techniques for using ultrasound to monitor corrosion, other, innovative, methods for corrosion sensing have been proposed using this principle as well. For example, some researchers have demonstrated that acoustic waves generated and received by fiber optic means can be used in a detector for monitoring corrosion in metal buried in concrete [15]. This device uses the fact that acoustic waves are blocked by corrosion products and this change in incident intensity across the rebar where fiber optic sensors have been embedded can be detected. This is an interesting concept but probably not very practical because of the need to bury sensors in the concrete. Another way to inspect metal buried in concrete for corrosion is by using guided wave techniques wherein ultrasonic energy is sent down the metal and mechanical defects detecting by electromagnetic acoustic transducers [16].

3.4.2 Radiography

This is a common modality for measuring cross-sectional profile of metal objects, such as sheet steel. Radiographic methods are used in inspecting pipes for their thickness and corrosion and erosion related defects. Gamma rays and high energy

X-rays are used to develop radiographs that can be used to determine wall thickness in pipes. Common sources are cobalt (Co-60) and iridium (Ir-192). Traditionally, film has been used to capture the image and post-process the information, but digital imaging techniques are now coming on board, much like in the medical sphere. Two techniques called tangential radiography and double wall radiography are standard. Radiography is superior to UT for pipe inspection if there is insulation around the pipe, which has to be removed for using ultrasonic techniques. The transverse technique depends on precise positioning of the source.

Generally, this is done by measuring the intensity of x-ray transmitted through the sample and thereby assessing the change in thickness. But there have been other methods such as scattering to evaluate thickness [17]. However, these seem to be more in the research realm.

3.4.3 Strain Measurements

Concrete, reinforced with pre-stressed iron bars (rebar), is the standard construction material for most large structures such as bridges. These must be inspected regularly for defects, one of which is the corrosion of buried rebar material. When buried steel corrodes, its volume increases, and many techniques have been developed for embedding strain gages in the hope of detecting this change in volume and hence the corresponding growth of corrosion products. Normal strain gauges or optical strain gauges can be used to accomplish this [18].

3.4.4 Acoustic Sensing

Acoustic emission corrosion measurements are based on measuring acoustic sound waves that are emitted during the growth of microscopic defects, such as stress corrosion cracks. The sensors can thus essentially be viewed as microphones, which are strategically positioned on structures. The sound waves are generated from mechanical stresses generated during pressure or temperature changes [19, 20].

3.4.5 Eddy Current

Eddy current techniques have long been used in NDT to measure defects and are the basis for proximity and vibration measurements in industrial applications. When applied to corrosion detection, they work on the principle of measuring skin thickness. In aerospace applications with aluminum skins, eddy current probes are used to measure thickness, much like ultrasonic probes. Eddy current probes do not need to make contact with the metal to get a reading, so it can be used in applications where UT may not work. Unlike radiographic techniques, this method of measuring thickness can be done from a single side.

The principle of this technique is to use a coil and an alternating current to generate an electromagnetic field, which is allowed to interact with the metal. The presence of the metal changes the effective impedance of the signal which can be measured. Defects on the metal surface or close under the surface, change the current flow, which has the same effect on the detected signal. The same principle works to measure thickness, since the other side of the metal structure can be considered to be a defect with infinite extent. Eddy current testing (ECT) can be done with continuous or with pulsed eddy current (PEC) signals. Because the latter essentially consists of multiple frequencies, more information can be garnered by the measurement of other features like rise time and final amplitude [21].

3.5 Sacrificial Sensors

A popular way of measuring corrosion is to measure the condition of a sacrificial sample of the same material as the structure, exposed to the same environment. The loss of mass or thickness or any other property correlated with corrosion can be used to indicate the presence of corrosion which could trigger a more thorough

Fig. 3.6 Corrosion test panels

inspection of the actual structure. Some adjustments need to be made to the estimates because the sample may not be loaded to the same extent as the actual structure, though fixtures can be used to load samples for improved environmental emulation and accelerated cracking. This can also be done within the detection algorithms by adjusting the thresholds at which inspections are called for.

Historically, sensing was done using coupons installed at or near the location where corrosion was being monitored. Periodically these calibrated coupons would be retrieved and measured to determine changes in dimensions, weight, etc. This would lead to an estimate of the corrosion rate at the given location. More locations would provide a more holistic picture of corrosion.

Corrosion test panels can be used to visually estimate the progression of corrosion over time. Or this can be assessed by weighing the samples to see the rate of loss of material. In Fig. 3.6 we see a set of corrosion test panels that have been exposed to the environment for different lengths of time, starting from the bottom left to the top right. This type of sensing is manual and visual, but at the same time cheap, effective, and commonplace.

3.6 Visual

Visual methods for estimating the extent of corrosion is quite common. Based on such measurements, guidelines exist for evaluation of a system's operational safety.

Visual Pit Counting

This is the most straightforward way of determining the number of corrosion pits on the surface. It can be done visually or under low magnification ($20\times$). As outlined in an ASTM standard [22] this counting can be done unaided or with the help of a movable plastic grid of 2 to 6-mm squares. This is done after thoroughly cleaning the surface. A micrometer or a depth gage can be used to determine the depth of accessible pits. With narrow pits, more sophisticated methods involving microscopy may be employed. Penetrant die may also be used to evaluate surface corrosion non-destructively, but this only gives an approximate measurement of the extent of damage. To get a more accurate sense of the extent of corrosion, samples of the metal need to be cut out and examined in the lab with more accurate microscopic instruments. These are destructive tests for corrosion.

With pit counting, standard charts exist for categorizing the density (A), the average size or extent (B), and the average depth or intensity (C) of the pits. The example shown in the ASTM standard says that "a typical rating might be A-3, B-2, C-3, representing a density of 5×10^4 pits/m^2, an average pit opening of 2.0 mm^2, and an average pit depth of 1.6 mm."

Other means of categorizing pitting intensity is to record the maximum pit depth or the average of the depths of a certain number of pits. Calculating the pitting factor, which is the ratio of the largest pit depth to the average depth, which works well for reasonably sized corrosion, is also an option.

3 Corrosion Sensing

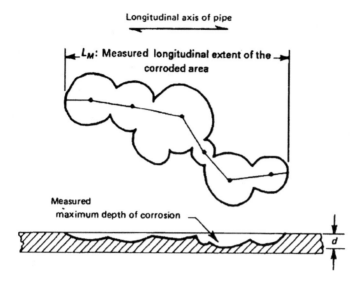

Fig. 3.7 Calculating parameters for ASME B31G analysis

Because pipes that carry fluids under pressure are susceptible to corrosion and, depending on the fluid, a disaster in the making if they leak or burst, standards have been developed to determine what is a safe amount of corrosion that can be tolerated. The American Society of Mechanical Engineers (ASME) has published a comprehensive guideline for determining how much surface corrosion can be tolerated before a pipe needs to be repaired [23]. This standard can be used by first determining the maximum depth of the corroded surface and finding its ratio to the nominal wall thickness. If this number is between 10 and 80% then the one is directed to measure the average longitudinal extent of the corroded area (see Fig. 3.7 for a cartoon version of this measurement).

The standard has tables that show the maximum value that this number can be to withstand a given pressure.

Visual detection of corrosion can be enhanced using automated image analysis, as, for example, described by Bonnin-Pascual and Ortiz [24]. The application they are looking at is the inspection of large areas such as the hulls of ships—spanning more 150 acres of metallic surfaces—where visual inspections are routinely conducted at enormous cost. They describe techniques that use image processing at the pixel level and using the hue and saturation values to distinguish corroded from non-corroded areas. This is done by training the algorithms over many example images. The authors also describe an Adaptive Boosting (AdaBoost) algorithm to do the classification, and their proposal is to use these as a first pass at corrosion sensing, before directing a more detailed visual inspection.

3.7 Optical

There are several optical methods for assessing corrosion. Surface corrosion can be sensed by profiling the surface using laser or other optical means, while more sophisticated fiber optical methods can be used to measure symptomatic changes due to corrosion such as reflectivity. Fiber optical cables afford the advantage of being able to sense structures over large distances since multiple sensors can be embedded on a single fiber.

3.7.1 Fiber Optic Methods

Fiber optic (FO) interrogation can be used to measure changes in the material translated into its optical properties. This is the basis for structural health monitoring of structures and a review article by Ye et al., gives a good overview of the techniques [25]. One way of monitoring reinforced concrete (RC) with embedded steel rebars is to realize that corrosion products are less dense, and hence they will expand during the oxidation process. This will apply pressure on the surrounding structure that can be measured either using standard strain gauges or using fiber optic sensors.

The most economical sensing systems are made by etching gratings at periodic lengths along a fiber and interrogating it using a laser light. These are called Fiber Bragg Gratings (FBG). If the fiber is attached securely to the material then any strain will result in the space between the FBG lengthening which can be detected by a change in the wavelength of the interference pattern made with the incident and the reflected light. This can be converted to intensity and calibrated to get a very accurate measurement of the strain; with a higher resolution than standard strain gauges. But this is not the only way FOs can be used in corrosion sensing.

One innovative sensor consists of an optical fiber that is mounted along a length of pipe, say underground where visual inspections are nearly impossible. The fiber can be many miles in length. At intervals along the fiber, 3–10% of the laser signal is split up into a branch fibers that are much shorter and have coatings at the end made of the same material as is being monitored. The number and length of the fiber and the side branches are limited only by the strength of the laser that can be employed economically and safely. The measurement is based on measuring the intensity of the return signal and the location is estimated by optical time domain reflectometry (OTDR). As the coating oxidizes due to the surrounding atmosphere, the intensity of the return signal drops so that the rate of corrosion can be measured. This is due to the fact that there is a gradual change in the optical reflectivity of the very thin coating material at the ends of the fibers. While innovative, it is not clear to what extent this has been commercialized [26].

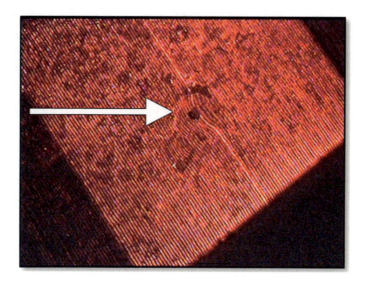

Fig. 3.8 Pit found and measured by laser profiling

3.7.2 Laser Profilometry

Laser profilometry is an optoelectronic technique to detect surface variations on a component or structure. By applying a laser light to the surface and collecting light reflected off of the surface, a laser triangulation technique can be sued to determine the distance from the laser to the surface based on the angle and amplitude at which the reflected light is received. By using this technique, surface defects caused by general corrosion and corrosion pitting can be detected and quantified (Fig. 3.8). Post-processing of images obtained with laser profilometry techniques can be used to determine material loss due to corrosion and can also support automated pit counting evaluations.

Generally used in non-destructive evaluation processes, laser profilometry can also be automated through the use of a scanning systems with built in two- or three-dimensional rastering systems. The outputs of laser profilometry systems can then be used to develop highly accurate measurements of surface profile and may be used to generate 3D models of a surface for comparisons before and after undergoing corrosion processes [27]. In the chapter on the use of analytics for detecting corrosion (Chap. 5), more details are provided on the use of advanced data analytic techniques to recognize and characterize corrosion from images.

3.7.3 White Light Interferometry

White light interferometry (also called coherence scanning interferometry for topographic measurements) is an optical means of measuring the surface imperfections such as those engendered by corrosion. Sub nano-meter resolutions can be achieved so that pit depths can be measured precisely. It is similar to profilometry but works on creating interference between split beams of white light and comparing reflected light from the surface to light reflected from a reference flat surface [28].

3.8 Other Sensing Modalities

While many corrosion sensing methodologies have been outlined in this chapter, additional mature and developmental techniques exist for the detection of corrosion and corrosive conditions. As corrosion sensing is an active field in the research and development community, many additional innovative methods have been created that have not yet been established and may not have sufficient in-field testing to prove out the technique just yet. Still others have shown promise in the laboratory, but do not rise to the level of active use in the field to date.

Proven measurement modalities including chemical sensing and physical parameter measurements can provide insights into the environmental conditions that will drive corrosion in a given scenario. Depending on the application, flow rate, pressure, and temperature can all play critical roles in corrosion processes. By monitoring these conditions, along with chemical analyses including pH, dissolved gas content (O_2, CO_2, H_2S, etc.), metal ion content, and microbial contaminants, it may be possible to quantify the corrosivity of an environment and infer corrosion rates for given materials. This indirect measurement technique can be particularly useful in submerged, flow loop, or pipeline applications.

Innovative solutions including multi-electrode arrays provide additional methods for detecting corrosion. As opposed to the double or triple electrode configuration standard for polarization resistance and potentiodynamic measurements, multi-electrode sensors leverage an array of metallic electrodes each electrically insulated from one another. By measuring current flow between each set of electrodes, corrosion rates can be determined. The multi-electrode configuration provides the advantage of improved simulations of bulk material corrosion over other sensing modalities, however additional instrumentation may be required for the multi-channel voltammetry measurements necessary for the technique [29]. Cathodic protection is a popular method for preventing, or minimizing, corrosion in structures (see details in Chap. 2). The multi-electrode corrosion sensor can be used to monitory the effectiveness of protection [30].

Other well-established damage detection and non-destructive evaluation (NDE) techniques have found new applications in corrosion monitoring. Direct current (DC) potential drop detection has a rich history in crack detection, particularly in laboratory applications. The technique has been modified using a four-point measurement probe to apply a DC potential to a surface, measure the resulting impedance, and track changes in that impedance as they relate to material corrosion. This measurement, of course, provides highly localized evaluations of corrosion as opposed to alternate "global' sensing approaches [31]. The LPR sensor has been mentioned above. Innovations in this technique include miniaturizing it and applying it to a wide variety of corrosion sensing domains [32].

As the corrosion sensing area continues to evolve, development of a truly comprehensive description of all available corrosion sensing techniques will be challenging. New techniques and analysis methods are becoming available on a regular basis and the field will likely continue to grow as technologies advance in the future.

3.9 Epilogue

Corrosion sensing plays a critical role in diagnostics and prognostics for materials degradation. Numerous techniques and technologies have been developed to enable detection and quantification of corrosion both in the laboratory and in the field. By analyzing data collected via electrochemical, environmental, or observational means, corrosion damage can be measured both directly and indirectly, resulting in actionable data for building and implementing diagnostic and prognostic tools for corrosion assessment.

References

1. Hinds G (1996) The electrochemistry of corrosion. Corrosion Doctors Publications, pp 4–5
2. Enos D, Scribner L (1997) The potentiodynamic polarization scan. Solartron Instruments, pp 1–14
3. Demo J, Andrews C, Friedersdorf F, Putic M (2011) Diagnostics and prognostics for aircraft structures using a wireless corrosion monitoring network. IEEEAC paper# 1181. IEEE, New York, NY
4. ANSI/NACE Standard TM0416-2016 (2016) Test method for monitoring atmospheric corrosion rate by electrochemical measurements
5. Holcomb G, Covino B (2001) State-of-the-art review of electrochemical noise sensors. Product of Department of Energy's Information Bridge: DOE Scientific and Technical Information
6. Reid SA, Eden DA (2001) US 6,264,824 B1. Assessment of corrosion. A method and apparatus for using EN to asses corrosion, preferably with skewness and kurtosis analysis using neural nets
7. Loto CA (2012) Electrochemical noise measurement technique in corrosion research. J Electrochem Soc 7:9248–9270
8. Kouril M, Prošek T, Scheffel B, Dubois F (2013) High-sensitivity electrical resistance sensors for indoor corrosion monitoring. Corrosion Eng Sci Technol 48:282–287. https://doi.org/10.1179/1743278212y.0000000074
9. Demo J, Friedersdorf F, Andrews C, Putic M (2012) Wireless corrosion monitoring for evaluation of aircraft structural health. IEEEAC paper # 1250. IEEE, New York, NY
10. ISO 9223:1992 (1992) Corrosion of metals and alloys—corrosivity of atmospheres—classification

11. Demo J, Kim M (2018) Corrosion management and asset tracking through improved airframe contaminant monitoring. NACE International
12. International Standards Organization (2012) SO 16809: non-destructive testing—ultrasonic thickness measurement. ISO16809
13. Ward B (2016) Ultrasonic inspection approaches for measuring corrosion wall loss in process piping. Quality Magazine
14. Strachan S, Fyedo M, Pellegrino B, Nugent M (2017) Non-intrusive ultrasonic corrosion-rate measurement in lieu of manual and intrusive methods. In: Proceedings of NACE corrosion conference and expo 2017
15. Du JC, Twumasi JO, Tang Q, Wu N, Yu T, Wang X (2018) Real time corrosion detection of rebar using embeddable fiber optic ultrasound sensor. In: Proceedings of SPIE 10598, Sensors and smart structures technologies for civil, mechanical, and aerospace systems 2018, https://doi.org/10.1117/12.2302901
16. Rose L (2002) A baseline and vision of ultrasonic guided wave inspection potential. ASME J Press Vessel Technol 124
17. Ong PS, Anderson WL, Cook BD, Subramanyan R (1994) A novel x-ray technique for inspection of steel pipes. J Nondestruct Eval 13/4
18. Almubaied O, Chai HK, Islam MdR, Lim KS, Tan CG (2017) Monitoring corrosion process of reinforced concrete structure using FBG strain sensor. IEEE Tran Inst Meas 66(8)
19. Zaki A, Chai HK, Aggelis DG, Alver N (2015) Non-destructive evaluation for corrosion monitoring in concrete: a review and capability of acoustic emission technique. Sensors 15(8): 19069–19101
20. Corrosion Doctor (2019) Acoustic Emissions (AE). https://corrosion-doctors.org/MonitorBasics/ae.htm. Retrieved April 2019
21. Silva IC, Santos YTB, Batista LS, Farias CT (2014) Corrosion inspection using pulsed eddy current. In: Proceedings of 11th ECNDT, Prague
22. ASTM International (2005) Standard guide for examination and evaluation of pitting corrosion. G46-94
23. ASME (2012) Manual for determining the remaining strength of corroded pipelines. B31G
24. Bonnin-Pascual F, Ortiz A (2014) Corrosion detection for automated visual inspection. In: Aliofkhazraei M (ed) Developments in corrosion protection. IntechOpen, Rijeka
25. Ye XW, Su YH, Han JP (2014) Structural health monitoring of civil infrastructure using optical fiber sensing technology: a comprehensive review. Sci World J http://dx.doi.org/10.1155/2014/652329
26. Martins-Filho J, Fontana E (2009) Optical fibre sensor system for multipoint corrosion detection. In: Lethiern C (ed) Optical fibre, new development. Intech Open, London
27. Alvarez RB, Martin HJ, Horstemeyer MF, Chandler MQ, Williams N, Wang PT, Ruiz A (2010) Corrosion relationships as a function of time and surface roughness on a structural AE44 magnesium alloy. Corrosion Sci 52
28. Freischlad K (2010) Optical surface profiling: profilometer advances benefit surface analysis, film-thickness measurement. Laser Focus World 46(1)
29. Yang L, Sridhar N (2002) US 6,683,463 B2. Sensor array for electrochemical corrosion monitoring. In: A sensor array for measuring localized corrosion based on electrochemical reactions
30. Sun X (2004) Online monitoring of corrosion under cathodic protection conditions utilizing coupled multielectrode sensors. Paper 04094, Corrosion 2004
31. Sposito G, Crawley P, Nagy PB (2010) Potential drop mapping for the monitoring of corrosion or erosion. NDT&E Intl 43
32. Connolly RJ, Brown D, Darr D, Morse J, Laskowski B (2013) Corrosion detection on buried transmission pipelines with micro-linear polarization resistance sensors. International workshop on smart materials, structures NDT in Canada, 2013 conference & NDT for the energy industry. Calgary

Chapter 4
Corrosion Prevention

Michael Casey Jones

Abstract Corrosion is a natural process, and thus, preservation of a material's properties by preventing interaction of that material with its environment is an essential component of any engineering design consideration. This chapter will demonstrate how the three primary protective coatings, organic, inorganic, and metallic coatings, are used within industry to prevent initiation of corrosion and slow the progression of corrosion upon its commencement. Corrosion is a continuous process, and thus engineers must use discretion in selecting proper coatings based on the environment that a material and its coatings will endure throughout the system's expected life cycle. Engineers must also use also understand the importance of design in preventing and controlling corrosion. Since corrosion is analogous to cancer throughout the manufacturing industry, the importance of preventing corrosion through proper coating and material selection is strongly emphasized over controlling corrosion once it's surfaced.

4.1 Introduction to Corrosion Prevention

The concept of corrosion maintenance is typically broken down into two areas: corrosion prevention and corrosion control. The latter concerns the process of minimizing corrosion once it has initiated via chemical reaction on a substrate with its environment. This chapter deals with the former, corrosion prevention, and thus will go into detail of the various mechanisms by which corrosion is prevented.

Corrosion prevention nearly always involves protecting a substrate via isolation from its environment and/or protecting the substrate by using compounds that are less noble, i.e., more anodic, than the substrate being protected. A common example of this that combines both of these protective measures would be the use of

---

M. C. Jones (✉)
USAF AFMC AFLCMC/EZP Product Support Engineering Division,
AFLCMC/EZPT-CPCO, Corrosion Prevention and Control Office,
325 Richard Ray Blvd, Bldg 165, Robins AFB, GA 31098-1639, USA
e-mail: mcjones60@gmail.com

© Springer Nature Switzerland AG 2020
G. Vachtsevanos et al. (eds.), *Corrosion Processes*, Structural Integrity 13,
https://doi.org/10.1007/978-3-030-32831-3_4

a Zn-rich primer on a steel substrate and overcoating with a polyurethane topcoat. The polyurethane topcoat serves as a barrier against the environment with tight crosslinks in the chemistry of the coating. Eventually, due to normal wear and tear via scratching, or degradation due to environmental factors, such as thermal shock, UV exposure, etc., the barrier properties of the topcoat will fail. Much work is being funded in the area of determining breakdown mechanisms of topcoats and primers via the Strategic Environmental Research and Development Program (SERDP) and Environmental Security and Technology Certification Program (ESTCP) [1, 2].

Once the topcoat has been compromised, the zinc in the epoxy primer reacts with an electrolyte, most commonly water via surface humidity, to sacrificially corrode so that the steel substrate is protected. The reaction of the zinc in the epoxy primer with its environment follows the mechanisms in Eqs. 4.1 and 4.2:

$$2H_2O \rightarrow 2OH^- + H_2 \tag{4.1}$$

$$Zn^{2+} + 2e^- + 2OH^- \rightarrow Zn(OH)_2 + 2e^- \tag{4.2}$$

Because the zinc is much more stable in its ionized state when in contact with iron, it more readily donates electrons in an electrolyte. The now positively charged zinc ion is thus going to react with the negatively charged hydroxide ions to reach its more stable, corroded state. Depending on the composition of the electrolyte, other corroded zinc products can be formed. These different electrolytes would also drive different reaction kinetics, as is commonly seen in industrial areas due to the reaction of sulfur dioxide (SO_2) with water to form sulfuric acid (acid rain), as well as highly salt-laden environments where sodium chloride (NaCl) readily dissociates into its ionized components of Na^+ and Cl^-. This form of corrosion prevention would be readily active on galvanized materials, a form of metallic coating which will be discussed later, such as hot-dipped galvanized nails.

While the aforementioned example is a common mechanism of corrosion prevention, there are other forms of corrosion prevention that we will explore in this chapter. Several forms of corrosion prevention involve the use of compounds that are considered toxic, such as cadmium and hexavalent chromium. While these compounds are known carcinogens, research in the area of replacing these has proven difficult because they prevent corrosion very well, thus making it difficult to find an equivalent replacement. Other forms of corrosion prevention, such as engineering design considerations, corrosion sensing, and corrosion preventative compound (CPC) will be explored later throughout this chapter.

Since a corroded state is a lower energy, equilibrium state relative to the originally manufactured form, corrosion is inevitable. Figure 4.1 shows a theoretical representation of why this is the case from a thermodynamic perspective. Energy is input into the process to manufacture a given alloy. The required energy input to achieve this state is represented by E_a, the activation energy. Because the Gibbs free energy, ΔG, required to go from the optimized alloyed stated to its corroded state is negative, the corroded state is an equilibrium state that the alloy will assume without corrosion preventative measures in place.

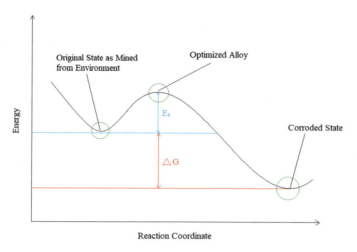

Fig. 4.1 Depiction of thermodynamically stable corroded state

However, robust corrosion prevention practices, especially in harsh environments, are almost always going to be more cost effective than corrective maintenance actions. While some forms of corrosion prevention may costly upfront, dismantling equipment to replace severely corroded parts will be time consuming and labor intensive, and thus more expensive than an active corrosion prevention program. In the following section, we will explore the many different coating options available for prevention of corrosion.

4.2 Coatings for Corrosion Prevention

Coatings for corrosion prevention can be summarized into three basic forms: organic coatings, inorganic coatings, and metallic coatings. Organic coatings for corrosion prevention are most commonly paints; however, other similar coatings, such as powder coatings, stains, varnishes, lacquers, and resins would also be considered organic coatings. Inorganic coatings are more commonly used in the pretreatment of materials, and examples of inorganic coatings would be anodization, conversion coatings, enamels, and ceramics. Metallic coatings are also commonly used as a form of corrosion prevention and control. Examples of metallic coatings would be cladding, electroplating, and thermal coatings.

In some cases, all three types of coatings may be used to prevent corrosion. For example, in the aerospace industry, typical aluminum alloy panels, such as 2024-T3 or 7075-T6 are coated clad with a thin layer of pure aluminum after manufacture. After this, the panels may then be anodized using chromic acid or thin film sulfuric acid anodization processes. Next, an adhesion promoter or chemical conversion coating (CCC) may be used to pretreat the substrate for organic coatings. After pretreatment has taken place, a metal rich epoxy primer would be applied, which may contain hexavalent chromium, lithium, magnesium, or some other inhibitor that protects against corrosion. Finally, a polyurethane topcoat is placed on top of the primer where it provides a barrier between the environment and the aforementioned corrosion prevention mechanisms. Despite all of these efforts, corrosion remains a multibillion dollar cost to industry and governments. The next section of this chapter will go into more detail on the various forms of coatings that can be used to prevent corrosion from occurring on a substrate intended to be protected from its environment.

4.3 Organic Coatings

Paints are the most common form of corrosion prevention. Coating substrates with paint is relatively cheap, and if the proper coating is selected for an asset's environmental exposure and usage conditions, it can also be very effective. Many different chemistries of paints are manufactured by major coating manufacturers. Some paint chemistries are known for better environmental protection, such as polyurethanes, which makes this a common selection as a topcoat. Other paint chemistries, such as epoxies, are better known for their adhesion properties, which make this a more common selection for a primer. Some of the more common coatings available are alkyds, epoxies, vinyls, rust converters, polyureas, and polyurethanes [3]. While all of these paints have different chemistries, paints generally have the same basic three components: solvents, binders (or resins), and pigments [4]. The solvent can be waterborne or solvent-borne (oil-based). The binder is the crosslinking chemistry that holds everything together, commonly consisting of materials in the isocyanate functional group as shown in Fig. 4.2.

Finally, the pigments are the additives, such as the colors, UV and corrosion inhibitors, biocides to prevent microbial growth, and so on.

When applying paints, common terminologies used in their application in manufacturers' technical data sheets (TDS) are wet film thickness (WFT) and dry film thickness (DFT). Common applications of paint include spraying, rolling, brushing, or immersion in a tank. Immediately after the coating has been applied, the WFT is the amount of coating that is on the surface, typically measure in mils or thousandths of an inch. As the paint goes through its curing process, the solvents in the paint will evaporate and the solids of the paint, the pigment and the resin, will remain behind to form the DFT. A manufacturer's TDS will dictate the proper DFT

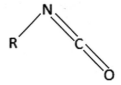

Fig. 4.2 Isocyanate functional group

4 Corrosion Prevention

to be applied. Measuring DFT using a WFT gauge can easily be accomplished by simply multiplying the WFT by the % solids content of the paint to give DFT. Average DFT can be calculated as shown below in Eq. 4.3

$$DFT(mils) = \frac{125 \times Volume\ of\ coating\ applied(in^3) \times \%\ solids}{surface\ area\ covered(ft^2) \times 18}$$

It's critical when considering paint application as a form of corrosion prevention to apply the in accordance with the manufacturers' TDS. Common considerations of paint application include, but are not limited to, temperature, relative humidity, cleanliness of the substrate, use of conversion coatings or adhesion promoters for surface preparation, time to mix the coating (induction time), thickness of the coating, time to apply the coating after mixing (pot life), and time between coating applications. Failure to follow these requirements in the TDS will invariably result in inferior performance of the paint.

While all efforts can be made to prevent coating failures from occurring, there are a multitude of various issues that a technician may experience while applying a coating. Table 4.1 shows a list of some of the most common coatings failures when applying paints [5].

Table 4.1 Common coating defects with causes alongside prevention and correction methodologies

Description	Causes	Remedies
Runs/sags: Characterized by an excess flow of paint	Spray gun too close, too much solvent/thinner, application of excess paint	Brush out excess paint before paint cure; if after paint cure, sand and reapply paint
Orange peel: Small indentures resembling the peel of an orange	Paint is too viscous, air pressure in spray gun too low, solvent evaporation too fast	Brush out excess paint before paint cure; if after paint cure, sand and reapply paint
Fish eyes: Separation or pulling apart wet film to expose undercoat or substrate	Application over dirty surface or incompatible surface. Applying outside of appropriate temperature range	Clean surface by abrasive blasting or sanding and apply a fresh coat
Lapping: Appearance of streaks from application of same coat	Coating drying too fast via fast-evaporating solvent or surface too hot	Change solvent for slower evaporation. Apply coating on top of itself prior to cure by applying in smaller areas
Pinholes/holidays: Appearance of missing coating	Not applying enough paint	Apply coating in slower passes to achieve a full coating; sand and repaint if cured
Blistering: Appearance of paint lifting off of the surface	Applying to an unclean or too hot of a surface	Remove paint via sanding, clean surface, and reapply at appropriate temperature range

These issues can happen due to defects within the paint, improper storage of the paint, use beyond the shelf life specified within the TDS, inexperienced technicians applying the coatings, and applying outside of the recommended environmental conditions specified in the TDS.

Another common form of organic coatings is powder coatings. Powder coats are unique in that they are applied electrostatically, where the powder is charged oppositely relative to the part that is being coated, thus causing the powder to stick to the substrate. After applying the powder to the surface, the powder is then cured in an oven at different temperatures and for different times in accordance with the manufacturer's TDS. As part of the alloying process, multiple different tempering schemes can be used to heat treat the material to alter the structural properties. Because of this, it's common to see multiple different curing schemes that offer lower temperatures for longer durations to prevent changing the temper of the material. Low temperature powder coatings (LTPC) are becoming increasingly popular because they cure around 300°F or less, and thus concern over altering the temper is minimal.

Powder coatings are classified in one of two categories: thermoset or thermoplastic. Thermosets are far more common than thermoplastics, and as a result, are generally less expensive. The primary difference between thermosets and thermoplastics are the mechanism of cure. Thermosets undergo an irreversible, chemical change during their cure process, whereas thermoplastics undergo a physical change and the powders are essentially melted together. A compromised thermoset powder coating is commonly touched up using typical organic coatings, such as a polyurethane topcoat. A thermoplastic, however, because it's undergone a physical change, can simply be heated with a heat gun and melted back together. This makes thermoplastics more maintainer friendly despite the higher upfront cost. Thermoplastics are also form very strong adhesive bonds with the substrate, but neither thermoplastics nor thermosets typically have any corrosion inhibitor. While powder coatings may have good adhesion, developments are still being made to increase the ultraviolet (UV) resistance via solar exposure.

4.4 Inorganic Coatings

Inorganic coatings fall more within the area of modification of the metallic surface chemistry than what is typically thought of as a coating. By modifying the surface chemistry, the substrate may become more corrosion resistant and/or provide a better surface for coating adhesion. Common examples of inorganic coatings include anodization, conversion coatings, and phosphatizing. Anodization and conversion coatings are commonly used on aluminums, whereas phosphatizing is used on steel.

Anodization is more commonly performed on aluminum and aluminum alloys than other substrates. Materials in their passive form are more corrosion resistant than in their active form, and anodization involves intentional passivation of the surface of the substrate because it's naturally more resistant. The anodization process is a multistep, electrolytic process that involves running direct current through acidic solutions and immersing the metals in the solution. Although an old standard that hasn't been updated in over two decades, MIL-A-8625 serves as the industry standard for anodization of aluminum alloys. Multiple types of acids are found at the beginning of MIL-A-8625, including chromic and sulfuric acids. Other types of acid, however, such as boric sulfuric, phosphoric, or oxalic acids can be used. Multiple variables can be altered, such as acid concentration, dwell time, voltage, and temperature to produce different depths of acid etch, offering different levels of corrosion protection, hardness, and other properties of the metal. Thinner metals generally require the use of chromic acid or thin film sulfuric acid, especially if the metal is under any fatigue loading. The reason is that these types produce lower etching depths and result in less of a fatigue knockdown as opposed to standard sulfuric acid. However, sulfuric acid is the most commonly used acid because it's much cheaper than chromic and some of the other acids mentioned.

While different parameters can be used to achieve different results, anodizing typically involves the followings processes: pretreatment, rinsing, etching, anodizing, and sealing. Pretreatment involves a cleaning process used to remove contaminants in a heated, alkaline solution. Rinsing in deionized water follows each of the steps to remove excess solution prior to going to the next step. After rinsing the pretreatment off, the substrate is dipped in a sodium hydroxide solution to etch the surface and prepare it for anodization. This etch increases the surface area, allowing more area for reaction of the acidic solution used in the anodization process to form passivated oxides. The etched metal is then rinsed and dipped in the anodization tank, where the substrate being dipped is negatively charged, hence the name anodization, relative to the acidic solution in the tank. This process generally produces an oxide layer 2–25 µm in thickness. Finally, the substrate is sealed using boiling water or a chemical, such as nickel acetate. This process is critical to prevent electrolytic intrusion, and therefore corrosion resistance [6].

Chemical conversion coatings are commonly used in industries to convert the surface from an active to a more passive state. Aluminum and magnesium are commonly conversion coated with chromate conversion coating. Since hexavalent chromium is highly toxic, non-hexavalent formations, such as trivalent chromium compounds, are being substituted in many chromate conversion coating processes. Not only do conversion coatings offer additional corrosion protection to the substrate, but they also modify the substrate to promote adhesion of subsequent coatings. MIL-DTL-5541 is the most commonly referred to material specification for chemical conversion coatings, with MIL-DTL-81706 being the closely associated process specification used for application of MIL-DTL-5541 conversion coatings. Upon looking at these specifications, one will notice there are multiple formulations and application methods that can be used for application of chemical conversion coatings.

4.5 Metallic Coatings

Use of a metallic coating is often part of the overall coating stackup for various applications. A metallic coating, as the name would indicate, is the use of a metal without any organic binders or resins. Metallic coatings typically take advantage of galvanic potential differences to offer protection to the underlying substrates. Application of the metallic coating can take place in one of multiple ways, including using electroplating, cladding, metallization/thermal spray, or dipping the substrate in molten material.

Electroplating is commonly used to create a barrier between the substrate and its environment through the use of a more noble and abrasion resistant coating. The coating can be created by immersing a cathodically charged substrate into a solution of anodically charged metallic particles. This process creates an electrical gradient by which the metallic particles are deposited on the substrate through electrical affinity. Various properties of the coating can be achieved by changing voltage, temperature, and dwell time in the tank. One of the most common examples of this include the use of zinc plate onto steel fasteners and brackets. This is a common treatment on ferrous substrates that is moderately effective if placed in a fairly mild environment. Another common example is the use of cadmium in the aerospace industry. While cadmium is considered highly toxic, it's also been shown to be highly effective for steel substrates. Some work has been focused in the area of the use of low hydrogen embrittlement zinc-nickel (LHE Zn-Ni) coatings as a replacement to cadmium. These can be applied through immersion, as previously discussed, or barrel plating. Barrel plating involves rotating parts in a barrel containing an electrolytic solution of metallic ions. Just as with immersion coating, charge differences drive the zinc-nickel particles to adhere to the substrate. Hydrogen embrittlement, which is classified as its own form of corrosion, is a common cause of failure of electroplated coatings. Aqueous solutions are commonly used for the electrolytic bath, and the presence of hydrogen gas in the solution can drive hydrogen gas into the substrate. This a common problem with high strength steels. A final method of electroplating, brush plating, is used by cathodically charging the substrate and brushing on the oppositely charged solution. This is more commonly used as a touchup process rather than a full coating. Many other metals can be applied via electroplating, such as copper, chromium, nickel, and so on.

Immersion in molten material is another common form of metallic coating. The most common form of this would be hot dip galvanizing. Hot dip galvanization is most commonly applied to steels to offer cathodic protection. The thickness of the coating obtained via hot dip galvanization is much thicker than would be achieved by zinc plating, and thus, offers superior corrosion protection. Thicker layers of zinc offer larger amounts of the anodically charged zinc relative to steel. Because this form of protection is fairly cheap and easy to apply, it's very commonly used as a protective coating on substrates.

Fig. 4.3 US coin clad with copper and nickel

Cladding is another metallic coating that can offer corrosion protection. Many of the coins used today, such as a U.S. dime or quarter, are examples of cladded materials. This can easily be observed by looking at the side of a coin, such as shown in Fig. 4.3.

Cladding can be accomplished via various mechanisms, such as rolling, extrusion, or welding. The basic principle remains that same, and heat and pressure are used to join the two metals. Manufacturers of aircraft commonly apply pure aluminum to 2024-T3 or 7075-T6 aluminum alloys to offer another layer of barrier protection. The impurities found in these aluminum alloys, such as copper and zinc, are added to increase structural properties, but often times serve as a location for pit formation due to highly localized galvanic corrosion. Pure aluminum doesn't readily corrode, hence its frequent use in canning in the food and beverage industry. Thus, applying a layer of pure aluminum to the surface provides an additional layer of corrosion protection.

Metallization, commonly referred to as thermal spray, is another metallic coating commonly used for corrosion protection. This form of protection can be cost prohibitive in some industries, however, it's been shown to exhibit exceptional corrosion protection in severely corrosive environments. The metal spray can be applying in multiple ways, by heating the material to be applied above its melting point and pneumatically spraying the molten material on. A common form of application is metal wire arc spray. In this form of metallization, two wires of opposite charge contact one another. Upon contact, an arc and melting of the wires occurs, and a high pressure air supply blows the molten material onto the substrate. Many different metals can be used as the spray, or even different combinations of alloys can be used. The United States Air Force authorizes the use of 85% zinc and 15% aluminum wire for protection of steel substrates. One very important consideration for the use of thermal spray is the temper of the substrate, and determining whether it will be impacted by applying hot, molten material. Of utmost importance is achieving a proper surface profile for the material to adhere properly. Various blast medias often need to be used to achieve a 2–4 mil angular surface profile in order for the thermal spray to adhere effectively.

4.6 Engineering Design Considerations and Coating Selection

The most effective way to prevent corrosion is through proper design. While nearly all materials are going to exhibit an adverse reaction with their environment, proper design can prevent corrosion from occurring. There are many considerations when purchasing new equipment, such as physical properties, appearance, cost, and so on, so corrosion can often be overlooked in the design phase. This is unfortunate and results in many unnecessary costs over the life cycle of an asset that could have been avoided with proper design. Corrosion should be considered as a cancer; thus, it's best to avoid it than to control it after it's started. Furthermore, preventative maintenance and measures, such as proper material and geometry selections or coating application, is almost always cheaper than corrective maintenance. Preventative maintenance involves the prevention of corrosion, whereas corrective maintenance often involves structural replacement or total condemnation of the asset. This section will consider the considerations that should be accounted for during the design of an asset.

Corrosion cannot occur without an electrolyte. Understanding this fundamental principle of corrosion science can't be understated. An electrolyte can present itself in the form of any liquid that's capable of transmitting electrons; thus, water would be the most common electrolyte that lends itself to corrosion. It's therefore critical that an asset be designed to minimize or eliminate fluid entrapment. This removes electrolyte necessary for corrosion to occur. Proper sealing, drain paths, and geometry are all essential considerations when designing an asset to prevent fluid entrapment. Flat surfaces should be minimized, but if necessary, should have adequate drainage to minimize the presence of the electrolyte.

Operating environment must always be considered during design of an asset. Important questions to ask include, what is the proximity to the coast, what is the climate, will the asset be in contact with corrosive chemicals (battery acid, deicing fluids, salt on streets, etc.), will the asset be near industrial pollutants, does the asset operate in extremely hot or cold temperatures, will the asset be primarily indoors or outdoors, and so on. These considerations, and an understanding of how they influence corrosion, will play an important role in material and coating selection. When designing an asset, the first consideration is usually the structural properties the material must possess to adequately serve its purpose. Typically, a multitude of materials can work; however, cost, appearance, weight, and corrosion resistance will inevitably be used to down select a product. For assets commonly used indoors that aren't exposed to abrasive conditions, cost and appearance will bias the decision. For assets used outdoors in inland conditions, organic coatings consisting of a pretreatment, epoxy primer, and polyurethane topcoat will typically suffice, but now corrosion resistance has some consideration. For assets that are going to be used in severe conditions, such as a marine environment or near industrial facilities, much more robust protection is needed and now corrosion resistance becomes crucial.

Highly corrosive environments commonly warrant the use of inorganic or metallic coatings in conjunction with organic coatings. Harsh environments are more typically corrosive because they have much longer times of wetness due to prolonged exposure to moisture. Thus, the electrolytic presence necessary for corrosion to occur is introduced. Furthermore, severely corrosive environments commonly have electrolytes will lower resistances, or in other words, increase conductivities. Higher conductivities lend themselves to more rapid rates of corrosion. Various factors can increase conductivity, such as presence of salts, acids, or other pollutants. Highly industrial areas have many sulfur dioxide (SO_2) emissions and thus react readily with moisture in the air (H_2O) to form sulfuric acid via the mechanisms in Eqs. 4.3 and 4.4.

$$2SO_2 + O_2 \rightarrow 2SO_3 \qquad (4.3)$$

$$2SO_3 + 2H_2O \rightarrow 2H_2SO_4 \qquad (4.4)$$

Furthermore, nitric acid can be formed via reactions of nitrogen dioxide (NO_2) with water via the mechanisms in Eq. 4.5.

$$4NO_2 + O_2 + 2H_2O \rightarrow 2HNO_3 \qquad (4.5)$$

Since sulfuric and nitric acids can be readily and reversibly dissociate into ionic compounds, they are both good electrolytes and thus promote corrosion. This can be seen in Eqs. 4.6 and 4.7.

$$2H_2SO_4 \leftrightarrow H_3SO_4^+ + HSO_4^- \qquad (4.6)$$

$$2HNO_3 \leftrightarrow NO_2^+ + NO_3^- + H_2O \qquad (4.7)$$

Another consideration for design is galvanic corrosion through interaction of dissimilar metals. This form of corrosion can easily be induced by changing the current flow properties of a metal via contact with another, more noble metal. Table 4.2 [7] is a common listing of corrosion propensities of various metals in sea water based on their electrical potentials. During design, the general rule is that the greater the potential differences between two metals in contact with one another in the presence of an electrolyte, the greater the rate of galvanic corrosion of the anode. This is not, however, always the case as the current flow is the dominating factor that influences corrosion. For example, quicker rates of galvanic corrosion of aluminum alloys are observed when coupled with stainless steel than when coupled with titanium, albeit Table 4.2 indicates greater galvanic corrosion rates of aluminum alloys would occur when coupled with titanium. MIL-STD-889, one of the governing documents of galvanic corrosion due to dissimilar metal contact, is undergoing revision because of differences in galvanic corrosion rates than potentials would suggest.

Table 4.2 Galvanic series of metals in sea water

ANODIC (High Corrosion Potential)
Lithium
Magnesium Alloys
Zinc (plate)
Beryllium
Cadmium (plate)
Uranium (depleted)
Aluminum Alloys
Indium
Tin (plate)
Stainless Steel 430 (active)
Lead
1010 Steel
Cast Iron
Stainless Steel 410 (active)
Copper (plate)
Nickel (plate)
AM 350 (active)
Chromium (plate)
Stainless Steels 350, 310, 301, 304 (active)
Stainless Steels 430, 410 (passive)
Stainless Steel 13-8, 17-7PH (active)
Brass, yellow, Naval
Stainless Steel 316L (active)
Bronze 220
Copper 110
Stainless Steel 347 (active)
Copper-Nickel 715
Stainless Steel 202 (active)
Monel 400
Stainless Steel 201 (active)
Stainless Steels 321, 316 (active)
Stainless Steels 309 13-8 17-7 PH (passive)
Stainless Steels 304, 301, 321 (passive)
Stainless Steels 201, 31, 6L (passive)
Stainless Steel 286 (active)
AM355 (active)
Stainless Steel 202 (passive)
Carpenter 20 (passive)
AM355 (passive)
Titanium Alloys
AM350 (passive)
Silver
Palladium
Gold
Rhodium
Platinum
Carbon/Graphite
CATHODIC (Low corrosion potential)

Prevention of galvanic corrosion is most easily performed by designing this out and not using dissimilar metals. If it's mandatory that a galvanic couple be used for other design reasons, it's essential that the two metals be insulated from one another using a neutral material, such as a sealant, rubber gasket, plastic insert, etc.

4.7 Conclusion

Corrosion, despite simply being defined as degradation of a material through chemical interaction with its environment to revert to its natural state, is a very complicated issue. This complication arises from the different environments that materials can be exposed to. Because environments can be localized, and as small as a battery compartment area, or more broadly defined, such as warm, coastal environments, it's very difficult to capture all necessary design, coating, and material selection considerations. To further complicate matters, a material's environment is constantly changing through diurnal cycles, seasonal changes, presence of chemicals, and movement from one environment to another. For these reasons, corrosion will always be an enemy of materials, and thus, the need for corrosion prevention through protective coatings and design considerations will serve an important role in maximizing a system's life cycle while minimizing routine maintenance costs.

References

1. Jones MC; WP-201710. Corrosion monitoring system to reduce environmental burden and corrosion maintenance costs. https://serdp-estcp.org/Program-Areas/Weapons-Systems-and-Platforms/Surface-Engineering-and-Structural-Materials/WP-201710/(language)/eng-US
2. Friedersdorf F; WP19-1168. Predictive coating condition model for advanced asset management. https://serdp-estcp.org/Program-Areas/Weapons-Systems-and-Platforms/Surface-Engineering-and-Structural-Materials/Coatings/WP19-1168/(language)/eng-US
3. Roberge P (2008) Corrosion engineering: principles and practice. pp 604–608
4. Society for Protective Coatings (SSPC) Aerospace engineering coating application training, Unit 6. pp 1–3
5. Roberge P (2012) Handbook of corrosion engineering, 2nd edn. p 791
6. Roberge P (2012) Handbook of corrosion engineering, 2nd edn. pp 821–825
7. Secretary of the Air Force (2019) Cleaning and corrosion prevention and control, aerospace and non-aerospace equipment. 25 Aug 2019, pp 2–10

Chapter 5
Data Analytics for Corrosion Assessment

George Vachtsevanos

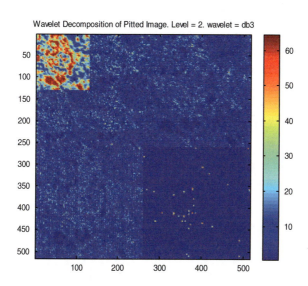

Abstract Accurate corrosion detection and prediction requires novel tools for processing surfaces/panels exposed to corrosive environments in order to detect early corrosion initiation. Data analytics offers a rich array of methods execute a sequence of tasks starting with pre-processing of raw images to improve the image to noise ratio, extracting features or useful information from pre-processed images, selecting the "best" features that are characteristic of corrosion evidence, and classifying the features for further processing. We take advantage of classical techniques for data mining but we introduce also new tools based on Deep Learning.

Keywords Corrosion data mining · Feature extraction · Feature selection · Classification · Wavelet features · Deep learned features · Sensor data fusion

---

G. Vachtsevanos (✉)
Georgia Tech, Atlanta, USA
e-mail: gjv@ece.gatech.edu

5.1 Introduction

It has been established that corrosion is one of the most important factors causing deterioration, loss of metal, and ultimately decrease of system performance and reliability in critical aerospace, industrial, manufacturing and transportation systems. Corrosion monitoring, data mining, accurate detection and quantification are recognized as key enabling technologies to reduce the impact of corrosion on the integrity of these assets. Accurate and reliable detection of corrosion initiation and propagation with specified false alarm rates requires novel tools and methods. Corrosion states take various forms starting with microstructure corrosion and ending with stress-induced cracking [1–3] (Fig. 5.1).

Generally speaking, corrosion starts in the form of pitting, owing to some surface chemical or physical heterogeneity, and then facilitated by the interaction of the corrosive environment fatigue cracks initiate from corrosion pitted areas and further grow into the scale that would lead to accelerated structure failure [4]. In order to effectively conduct structural corrosion health assessment, it is thus crucial to understand how corrosion initiates from the microstructure to the component level and how structure corrosion behaviors change as a result of varied environmental stress factors. Many research efforts have been reported in the past addressing this critical issue [5–8]. Traditionally, conventional ultrasonic and eddy current techniques have been used to precisely measure the thickness reduction in aircraft and other structures; there has been a number of undergoing research efforts

- **Uniform Corrosion**
  - Very high or very low pH
- **Galvanic Corrosion**
  - Dissimilar metals making electrical contact in a solution.
- **Micro-structure corrosion**
  - Pitting
    - Common denominator in almost all types of corrosion attack
    - May assume different shapes
    - Chlorides (Cl^-)
  - Intergranular corrosion
    - Grain-boundaries
    - Alloys / Constituents
  - Stress Corrosion Cracking
    - Tensile Stress

http://www.nace.org/Pitting-Corrosion/

Image showing a pit formation

Image showing intergranular corrosion

Fig. 5.1 Microstructure corrosion

using guided wave tomography techniques to screen large areas of complex structures for corrosion detection, localization [9] and defect depth mapping [10]. However, due to the nature of ultrasonic guided waves, this technique is vulnerable to environmental changes, especially to temperature variation and surface wetness occurrence [11], and the precision of corrosion defect depth reconstruction is restricted by sensor network layout, structure complexity, among others, which limits the scope of the field application. Thus, undeniably, well-recognized global corrosion measurements, such as material weight loss and wall thickness reduction, cannot offer an appropriate and trustworthy way to interpret the pitting corrosion due to its localized attack nature.

Figure 5.2 is a pictorial representation of the corrosion assessment technologies from corrosion monitoring to data mining, detection/prediction and assessment.

Besides, advanced corrosion health assessment systems require comprehensive quantitative information, which can be categorized into a variety of feature groups, such as corrosion morphology, texture, location, among others. It calls for the exploration of both new testing and data fusion methods from multiple testing

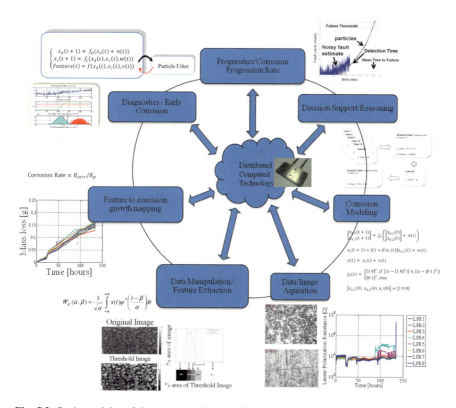

Fig. 5.2 Basic modules of the smart sensing modality

techniques. Forsyth and Komorowski [12] discussed how data fusion could combine the information from multiple NDE techniques into an integrated form for structural modeling. Several other studies have looked into different sensing technologies for corrosion health monitoring, including using a micro-linear polarization resistance, μLPR sensor [13, 14], and fiber optic sensors [15]. However, the existing research effort in combination with surface metrology and image processing is very limited. In parallel to the current corrosion sensing technology, there have been a number of corrosion modeling studies trying to numerically capture the processes of pitting corrosion initiation, pitting evolvement, pitting to cracking transition, and crack growth to fracture at the molecular level. However, currently there is no accepted quantitative model to take into consideration the effect of stress factors (e.g., salinity, temperature, humidity, pressure), although the effects of the above-mentioned stress factors have been widely discussed.

We address in this chapter analytical tools and methods to extract useful information from corroding surfaces that will be exploited eventually to assess the health state of critical aircraft, ships, and transportation systems, among others [16]. The architecture is set as a decision support system providing advisories to the operator/maintainer as to the health status of such assets subjected to corrosion and in need of corrective action.

5.2 Corrosion Data Mining-Feature Extraction and Selection

An important and essential component of the corrosion detection and interpretation architecture involves image/characterization data pre-processing and data mining aimed to extract useful and relevant information from raw data. Figure 5.3 depicts the architectural components of the pre-processing, feature extraction, selection and classification steps. The latter is detailed in the sequel.

Features are the foundation for the degradation/corrosion detection and interpretation scheme. Feature extraction and selection processes are optimized to extract only the information that is maximally correlated with the actual corrosion state. Appropriate performance metrics, such as correlation coefficients, Fisher's Discriminant Ratio (FDR), etc. are utilized to assist in the selection and validation processes. Figure 5.4 shows the overall data mining scheme. Image pre-processing, feature extraction and selection are highlighted leading to their utility in pitting corrosion detection, localization, quantification, and eventually prediction of corrosion states.

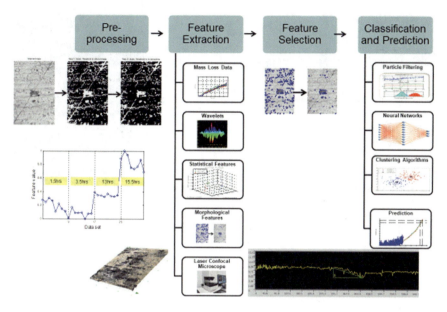

Fig. 5.3 Corrosion data pre-processing, feature extraction/selection and classification

Fig. 5.4 Corrosion data mining scheme

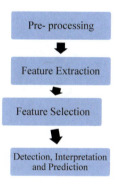

5.3 Image Pre-processing

Image/data pre-processing involves filtering and preparing the data for further processing. Figure 5.5 shows a typical sequence of pre-processing steps of corrosion images from surface metrology testing. In the first step, de-noising, discrete stationary wavelet transform (SWT) is applied, and then histogram equalization is performed for contrast enhancement followed by applying a threshold to identify the regions of interest in the image. In this framework, image processing techniques are utilized to pre-process the global test panel images as well as the local pitting area images, in preparation for the feature extraction step. First, globally, for each

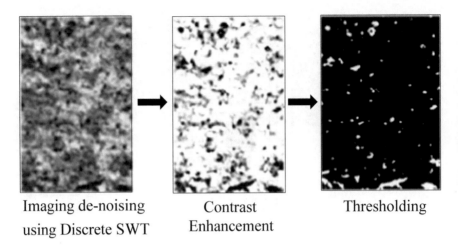

Fig. 5.5 Corrosion image pre-processing

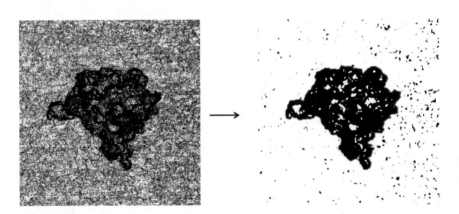

Fig. 5.6 Local pit identification via image processing. Left: Original localized pit image; Right: Pit identified from the background with the pit edge (in blue) identified by image processing algorithm

test panel used to demonstrate the algorithmic developments, successive 2D microscopic images were taken and stitched together to obtain the entire panel image. In the whole panel image pre-processing, the rivet-hole areas and artifacts (e.g., stencil-stamp marked numbers) were manually whitened so they would not be confused with corroded regions. In order to identify the pitting corrosion attacked areas, a 2D median filter was applied followed by thresholding (with a threshold of 0.2) to obtain at a binary image. Second, locally, each suspected pitting area is identified from the whole panel image, and a closer microscopy examination was conducted. An example of a local pit identification process is shown in Fig. 5.6. To identify the pit(s) from the background, the area of each object (i.e., black region representing corroded region) in the binary image was calculated. The sum of objects with the area larger than 50 pixels was defined as the total area of the pitting corroded regions. Note that the identification threshold of 50 pixels was set to avoid mistaking dark regions caused by the grain boundaries as pits.

5.4 Data Mining/Image Processing

An important and essential component of the corrosion detection architecture involves data/image pre-processing and data mining aimed to extract useful and relevant information from raw data. In the proposed architecture, the most important components supporting the implementation of the algorithm are feature extraction and diagnosis/prognosis models. Features are the foundation for "good" fault/corrosion detection algorithms. Feature extraction and selection processes are optimized to extract only information that is maximally correlated with the actual corrosion state. Appropriate performance metrics are defined to assist in the selection and validation processes. Image/data preprocessing involves filtering and preparing the data for further processing. Figure 5.7 shows a typical sequence of preprocessing steps for corrosion images.

Of particular interest to our theme is localized pitting and cracking, i.e. cracking initiating at points on the surface of a specimen (joints, fasteners, bolts, etc.). A metal surface (aluminum alloy, etc.) exposed to a corrosive environment may, under certain conditions experience attack at a number of isolated sites. If the total area of these sites is much smaller than the surface area then the part is said to be experiencing localized corrosion in Fig. 5.8. We exploit novel image processing tools/methods, in combination with other means (mass loss calculations) to identify

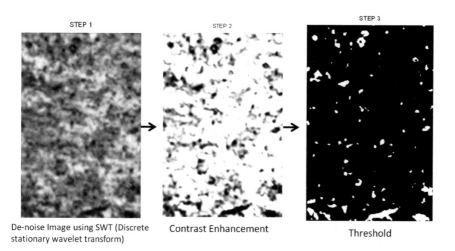

Fig. 5.7 Image preprocessing

Fig. 5.8 Left: LEXT OLS4000 3D Laser Measuring Microscope. Right: Bruker's Dektak® Stylus Profiler

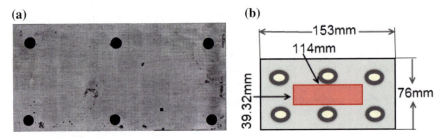

Fig. 5.9 a Whole plate imaging of AA 7075-T6 panel (150.63 × 73.87 mm, 108×). b Area the profilometer covers for the 3D map scan

features of interest to be used in the modeling task, since imaging of corroding surfaces offers a viable, robust and accurate means to assess the extent of localized corrosion (Fig. 5.6).

We use novel microscopy and profilometry image processing equipment in order to obtain images of corroded surfaces and extract from such images relevant information that assists in corrosion modeling, diagnostics and prognostics. In our testing, we are using a laser microscope and a stylus-based profilometer, as shown in Fig. 5.9.

The LEXT OLS4000 3D Laser Measuring Microscope is designed for nanometer level imaging, 3D surface characterization and roughness measurement. Magnification ranges from 108× to 17,280×. Typical 2D and 3D images are shown in Figs. 5.9a and 5.10, respectively.

Fig. 5.10 Pitted panel area (1278 × 2561 μm, 216×) 3D imaging and corresponding profile info (in μm) on uncoated AA 7075-T6 panel

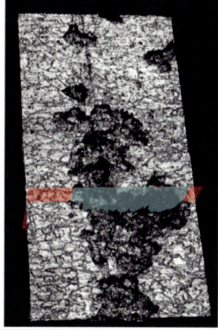

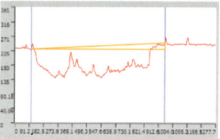

The Bruker's Dektak® Stylus Profiler is a traditional 2D contact profilometer. However, with the programmable map scan capability and the post-processing software, it allows for large area 3D topography coverage. The combination of the two imaging tools facilitates both the localized and global characterization of a corroded panel at various resolution scales. In summary:

1. Global characterization:
   - The laser microscope can provide large area 2D microscopy imaging as shown in Fig. 5.9a.
   - The stylus profilometer can provide large area 3D map scan imaging. A schematic of the area the profilometer covers for the 3D map scan for a typical panel is shown in Fig. 5.9b.

2. Local characterization:

- After the detection and localization, the laser microscope can provide a close look at the 3D topography of the analyzed surface. An example of pit profile measurement is shown in Fig. 5.6.

5.5 Feature Extraction and Selection

The images obtained through conventional NDI methods are not directly suitable for identification and quantification of damaged regions. Such images therefore need to be enhanced and segmented appropriately for further image analysis. Segmentation has been achieved using de-noising, contrast enhancement, and threshold techniques. Transform based features such as wavelet coefficients also can be used to quantify the extent of corrosion in the image. Neural networks were applied in the process of segmentation and quantification of damaged regions. Segmentation results show a good correspondence between the extracted regions and the actual damage on sample panels.

Several features may be extracted from corrosion images such as statistical features, transform based features (wavelet), texture features and morphological features. These features are outlined in Tables 5.1, 5.2, 5.3 and 5.4.

Figure 5.11 shows a set of example corrosion images that were taken with the LEXT OLS laser measuring microscope. The images were taken of an AA 7075-T6

Table 5.1 Statistical features

Feature name	Symbol	Formula	Comments
Mean pixel value	$\bar{x}$	$\frac{1}{MN}\sum_{m=1}^{M}\sum_{n=1}^{N} I(m,n)$	
Standard deviation of pixel values	std	$\sqrt{\frac{1}{MN}\sum_{m=1}^{M}\sum_{n=1}^{N}(I(m,n)-\bar{x})^2}$	
ith moment	m_i	$\frac{1}{MN}\sum_{m=1}^{M}\sum_{n=1}^{N}(I(m,n)-\bar{x})^i$	For i = 2 this is equivalent to the standard deviation squared
Entropy	e	$-\sum_{i=1}^{255} p(i)\log_2(p(i))$ Where p(i) is the probability of a pixel in the image having a value of i. This is calculated from the histogram of the image I	A scalar value representing the entropy of an intensity image. Entropy is a statistical measure of the randomness that can be used to characterize the texture of the input image

Table 5.2 Wavelet features

Feature name	Symbol	Formula
Total energy	E_k^i	$E_k^i = \frac{1}{MN} \sum_{m=1}^{M} \sum_{n=1}^{N} D_k^i(m,n)^2$ where k = H, V or D (corresponding to horizontal, vertical or diagonal subimage) and i is the level of the wavelet decomposition
Anisotropy	$Orian^i$	$Orian^i = \frac{1}{MN} \sum_{m=1}^{M} \sum_{n=1}^{N} Orian(m,n)$ where $Orian(m,n) = \frac{1}{E_{total}^i(m,n)} \sqrt{(D_H^i - D_V^i)^2 + (D_H^i - D_D^i)^2 + (D_V^i - D_D^i)^2}$ $E_{total}^i(m,n) = (D_H^i(m,n))^2 + (D_B^i(m,n))^2 + (D_D^i(m,n))^2$

Table 5.3 Texture features from gray level co-occurrence matrix (GLCM)

Feature name	Symbol	Formula	Comments
Contrast	c	$\sum_{i,j} |i-j|^2 p(i,j)$ where p(i, j) is the normalized gray level co-occurrence matrix	Returns a measure of the intensity contrast between a pixel and its neighbor over the whole image. Contrast is 0 for a constant image
Correlation	correl	$\sum_{i,j} \frac{(i-\mu_i)(j-\mu_j)p(i,j)}{\sigma_i \sigma_j}$ where p(i, j) is the normalized gray level co-occurrence matrix	Returns a measure of how correlated a pixel is to its neighbor over the whole image. Range = [−1 1]
Energy	E	$\sum_{i,j} p(i,j)^2$ where p(i, j) is the normalized gray level co-occurrence matrix	Returns the sum of the squared elements in the GLCM. Range = [0 1]. Energy is 1 for a constant image
Homogeneity	H	$\sum_{i,j} \frac{p(i,j)}{1+|i-j|}$ where p(i, j) is the normalized gray level co-occurrence matrix	Returns a value that measures the closeness of the distribution of elements in the GLCM to the GLCM diagonal. Range = [0 1]. Homogeneity is 1 for a diagonal GLCM

panel that has an exposure time of 286 h in a cyclic corrosion chamber running the ASTM G85-A5 test. In order to distinguish between images of different levels of corrosion a number of features can be extracted from the images. These features fall under the main categories of statistical features, transform based features (wavelet), texture features and morphological features. The statistical features are summarized in Table 5.1. The mean pixel value, standard deviation and entropy of the example corrosion images in Fig. 5.11 are plotted in Fig. 5.12.

Table 5.4 Morphological features

Feature name	Symbol	Formula	Comments
Roundness	r	$r = \frac{4\pi A}{p^2}$ where A is the area of the region and p is the perimeter of the region	If the region is a circle then r = 1. This is a good feature to distinguish between cracks and pits if segmentation is done correctly
Solidity	s	$S = \frac{Area}{Convex\,Area}$	Scalar specifying the proportion of the pixels in the convex hull that are also in the region
Eccentricity	ecc	$ecc = \sqrt{1 - \frac{b^2}{a^2}}$ for ellipse defined by: $\frac{x^2}{a^2} + \frac{y^2}{b^2} = 1$	Scalar that specifies the eccentricity of the ellipse that has the same second-moments as the region. An ellipse with ecc = 0 is a circle and an ellipse with ecc = 1 is a line segment
Major axis length	L_{Major}	$L_{Major} = \max(2a, 2b)$ for an ellipse defined by: $\frac{x^2}{a^2} + \frac{y^2}{b^2} = 1$	Scalar that specifies the major axis length of the ellipse that has the same second-moments as the region
Minor axis length	L_{Minor}	$L_{Minor} = \min(2a, 2b)$ for an ellipse defined by: $\frac{x^2}{a^2} + \frac{y^2}{b^2} = 1$	Scalar that specifies the minor axis length of the ellipse that has the same second-moments as the region
Area	A	$A = \#\,pixels\,in\,the\,region$	Area of the region
Mean intensity of the region	MI	$MI = \frac{1}{A} \sum_{(m,n) \in R} I(m,n)$ where R is the region of interest	Mean intensity of the region of interest in the image

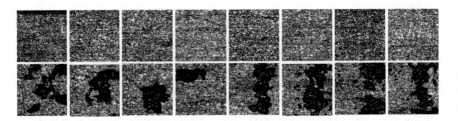

Fig. 5.11 Example corrosion images. Top row: low corrosion. Bottom row: high corrosion

Wavelet based features are widely used in the literature for Image processing/data mining applications. The wavelet domain takes advantage of the wavelet capability to resolve images in the spatial and frequency domains. The 2D wavet decomposition pursues the following steps:

5 Data Analytics for Corrosion Assessment

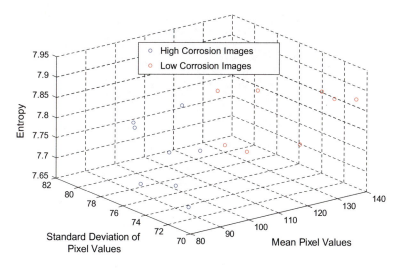

Fig. 5.12 Statistical features of example images shown in Fig. 5.1

Approximation	Horizontal Detail 2
	Horizontal Detail 1
Vertical Detail 2	Diagonal Detail 2
Vertical Detail 1	Diagonal Detail 1

- A wavelet transform of a 2D image, I, can be performed by applying a set of band and low pass filters H and L along the rows (x) and columns (y) of the image. The sub-images of the 2D wavelet transform are given by the following equations:

$$A = [L_x * [L_y * I]](x,y)$$
$$D_H = [L_x * [H_y * I]](x,y)$$
$$D_V = [H_x * [L_y * I]](x,y)$$
$$D_D = [H_x * [H_y * I]](x,y)$$

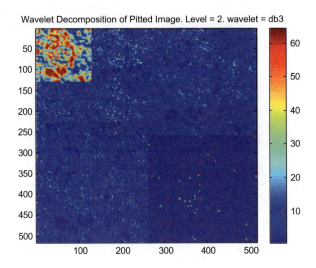

Fig. 5.13 2 level 2D wavelet decomposition

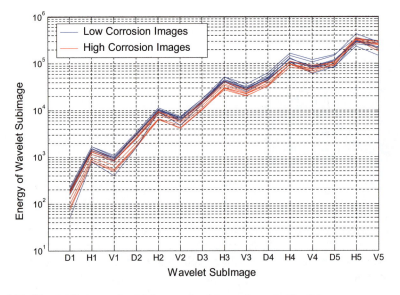

Fig. 5.14 Wavelet features of example corrosion images shown in Fig. 5.1

where * denotes convolution, A is the approximation image, D_H is the horizontal detail, D_V is the vertical detail, and D_D is the diagonal detail

- After the band and low pass filters are applied the 4 resulting sub-images are down sampled.

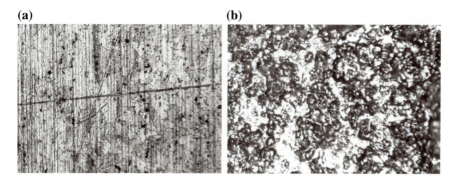

Fig. 5.15 Corrosion image processing

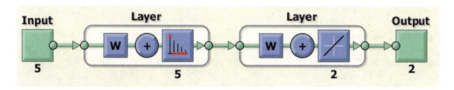

Fig. 5.16 Images from submersion test used to test wavelet feature extraction method. **a** Corresponds to "baseline" and **b** corresponds to "pitted"

- The sub-images at the next level of thwavelet transform are computed using the same operations on the approximation sub-image from the previous level.

Figure 5.13 shows an example of a 2-level wavelet decomposition of a corrosion image using a Daubechies wavelet (db3). Table 5.2 shows the wavelet features that are extracted at each wavelet decomposition level as outlined by Livens. At each level, k, a total energy feature, E_{total}^k, was calculated using the horizontal, vertical and diagonal sub-images at level k (D_H^k, D_V^k, and D_D^k respectively). Additionally the anisotropy of the energy (how much the energy differs with direction) was calculated. Figure 5.14 shows the wavelet energy features of the example corrosion images shown in Fig. 5.11. From the graph it is evident that the low corrosion images seem to have higher energy feature values at each level than the high corrosion images.

Wavelet Feature Extraction and Classification example:

- Step 1: For each 512 × 512 block perform 2D wavelet decomposition as shown in Fig. 5.13. 512 × 512 blocks were taken from images shown in Fig. 5.15. Figure 5.16 shows the processing steps.
- Step 2: Extract energy and Orian features from wavelet coefficients: $E_{total}^1, E_{total}^2, Orian^1, Orian^2, E_A$ (see Table 5.2).

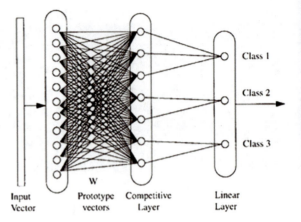

Fig. 5.17 Learning Vector Quantization (LVQ) neural network

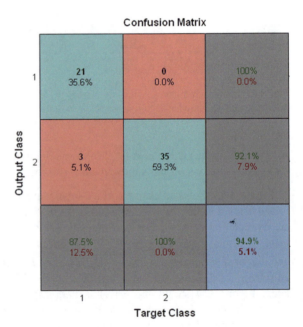

Fig. 5.18 Confusion matrix. Class 1 corresponds to pitted images and Class 2 corresponds to baseline image. The Target class is the actual class of the image output class is the predicted class of the image from the LVQ neural network

- Step 3: Train Learning Vector Quantization (LVQ) Neural Network (shown in Fig. 5.17) with 30 random samples.
- Step 4: Test LVQ with entire set of samples (results shown in Fig. 5.18). Only three pitted images were misclassified as a baseline image (Fig. 5.19).

Texture features such as contrast, correlation, energy and homogeneity can be calculated using the gray level co-occurrence matrix (GLCM) of an image (see Table 5.3). The (i, j) value of the GLCM of an image I has the value of how often a

5 Data Analytics for Corrosion Assessment

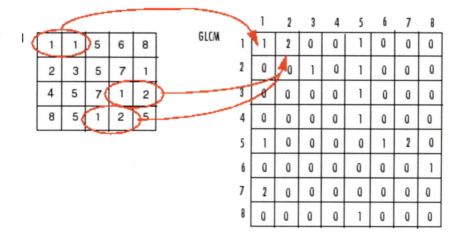

Fig. 5.19 GLCM calculation from image

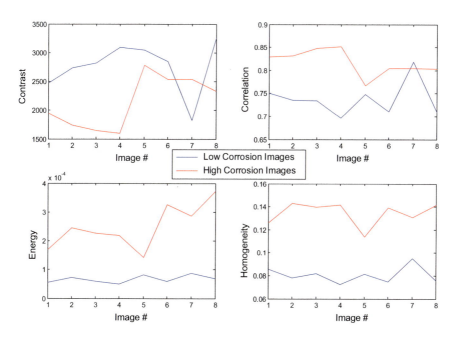

Fig. 5.20 Contrast, correlation, energy and homogeneity features of low and high corrosion images (Fig. 5.14)

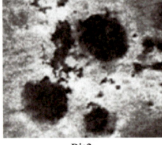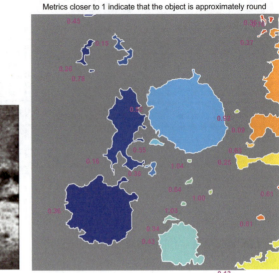

Fig. 5.21 Image on the right shows all the objects identified in the image 'Pit3' with the roundness metric in pink near the object

pixel value i occurs horizontally adjacent to a pixel with value j in the image I. Figure 5.19 shows a depiction of how to calculate the GLCM matrix from an image I. Example values of the GLCM features calculated using the low and high corrosion images (as shown in Fig. 5.11) are shown in Fig. 5.20. Note that the high corrosion images tend to have higher homogeneity, correlation and energy. This could be due to the large black areas that characterize the high corrosion images.

Table 5.4 displays a number of morphological features that can be extracted from a segmented image. Morphological features give information on the type of shapes in the images. Pits tend to be round objects while cracks are longer and tend to have an eccentricity close to 1. An example of Fig. 5.21 shows an example of the roundness feature calculated for an image of pits. These morphological features can be used to characterize the types of different corrosion states in one image (for example identifying a crack or pit in an image) (Fig. 5.22).

The area feature is used to calculate the percentage of corrosion on an entire panel. Figure 5.22 shows a panel from the March 12 2014 lap joint chamber test. The top image is the uncoated 7075-T6 aluminum panel that had an exposure time of 286 h. The bottom image is the binary image of the top image after applying a median filter. From the binary image the percent area of corrosion on the panel can be calculated.

Figure 5.23 shows the progression of corrosion during the March 12 2014 lap joint chamber test through the three images on the left and the graph of their corresponding percent area of corrosion on the right. One of the main disadvantages

5 Data Analytics for Corrosion Assessment

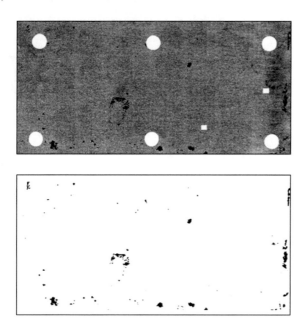

Fig. 5.22 Top: Original image with rivets and number removed. Bottom: Binary image after filtering. Black corresponds to corroded regions

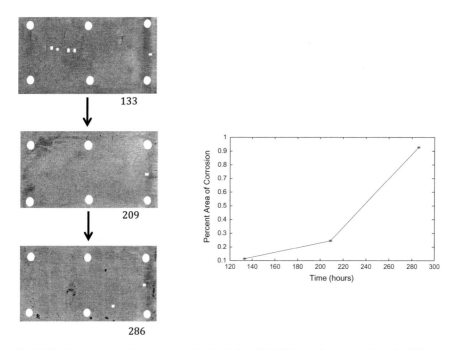

Fig. 5.23 Percent area of corrosion calculated for 7075-T6 aluminum panels with different exposure times

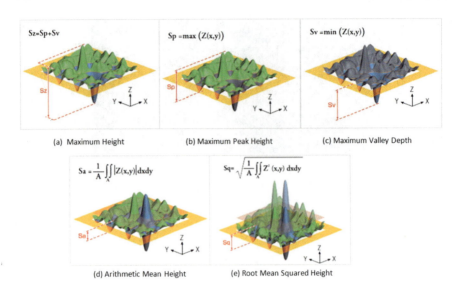

Fig. 5.24 Surface roughness features

Table 5.5 Surface roughness features

Name	Symbol	Equation	Figure
Maximum height	S_Z	$S_Z = S_P + S_V$	5.24a
Maximum peak height	S_P	$S_p = \max(Z(x,y))$	5.24b
Maximum valley depth	S_V	$S_V = \min(Z(x,y))$	5.24c
Arithmetic mean height	S_a	$S_a = \frac{1}{A}\int |Z(x,y)|dxdy$	5.24d
Root mean squared height	S_q	$S_q = \sqrt{\frac{1}{A}\int |Z(x,y)|^2 dxdy}$	5.24e
Skewness	S_{sk}	$S_{sk} = \frac{1}{S_q^3}\frac{1}{A}\int Z(x,y)^3 dxdy$	
Kurtosis	S_{ku}	$S_{ku} = \frac{1}{S_q^4}\frac{1}{A}\int |Z(x,y)|^4 dxdy$	

of using morphological features is that they are very sensitive to the segmentation algorithm that is used (Fig. 5.24).

In addition to features that can be extracted from the 2D corrosion images the LEXT OLS4000 3D Laser Measuring Microscope and Veeco Dektak 150 surface profilometer can measure a number of surface roughness and volume features. The surface roughness features are listed in Table 5.5. Note that Z(x, y) is the height of the panel measured over an area, A, of about 39.32 mm × 114 mm using the Veeco Dektak 150 surface profilometer.

Parameters related to the volume of the void portion and the material portion are defined as shown in the diagram in Fig. 5.25. 10 and 80% are default values of the heights for the boundaries among the valley section, core section, and peak section.

5 Data Analytics for Corrosion Assessment

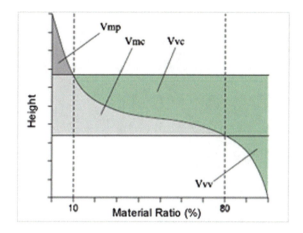

Fig. 5.25 Material ratio curve

Fig. 5.26 Two 1.0" × 1.0" AA 2024-T3 and AA 7074-T6 panels prepared for microscopic analysis

- V_{vv}: The void volume of the valley section, as calculated from the material ratio curve
- V_{vc}: The void volume of the core section, as calculated from the material ratio curve
- V_{mp}: The material volume of the peak section, as calculated from the material ratio curve
- V_{mc}: The material volume of the core section, as calculated from the material ratio curve.

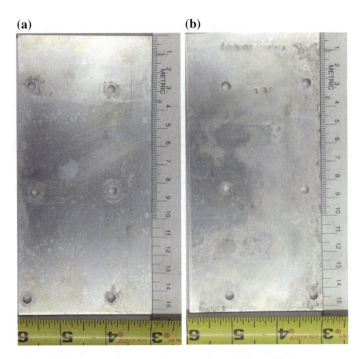

Fig. 5.27 Corrosion panel #1 of AA2024-T3. **a** Uncoated side and **b** coated side

Table 5.6 Summary of surface profile measurement of Panel #1, AA2024-T3 ("×" represents the measurement is close to a random rivet hole, while the blank represents being away to any rivet holes)

No.	1	2	3	4	5	6	7	8	9	10
Uncoated			×	×	×	×				
Coated					×	×		×		

Surface roughness and volume features measured using the LEXT OLS4000 3D Laser Measuring Microscope are shown in Figs. 5.28, 5.29 and 5.30. The features were measured from the following panels:

1. **Baseline sample of AA 2024-T3**: Surface profiles are measured including the 2D and 3D profile images and basic profile information, and surface roughness is calculated. Panel shown in Fig. 5.26.
2. **Corrosion Panel #1 of AA 2024-T3 from 2013 – jan 16 BAA-RIF lapjoint cct−10 test**: 10 measurement areas of 642×644 μm^2 were randomly selected (close/away from rivet holes) from both coated and uncoated sides for profile the measurement, and 3D surface roughness is calculated for each measurement. This panel is shown in Fig. 5.27; (Table 5.6).

5 Data Analytics for Corrosion Assessment

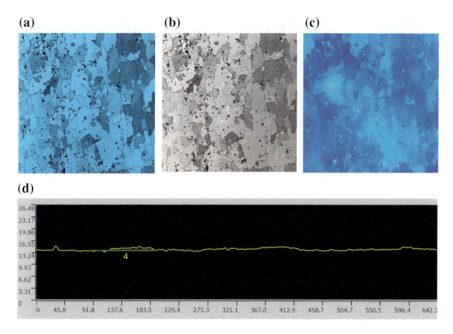

Fig. 5.28 Surface 2D measurement of the AA 2024-T3 baseline sample. **a** Laser intensity image. **b** Color image. **c** Height intensity image. **d** A cross-section profile

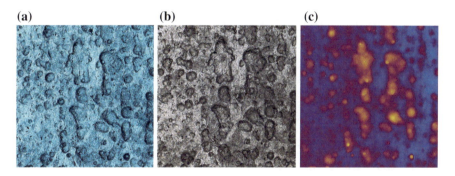

Fig. 5.29 Surface 2D measurement from panel #1 uncoated side, close to a rivet hole, AA 2024-T3. **a** Laser intensity image. **b** Color image. **c** Height intensity image

5.6 Baseline Profile Measuring Results

5.6.1 2D Profile Information

Corrosion Panel #1 Profile Measuring Results (Figs. 5.28 and 5.29)

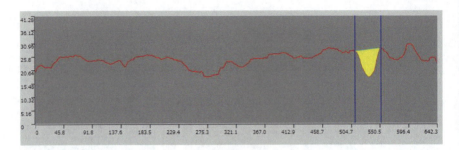

Fig. 5.30 A pit profile from Panel #1, uncoated side, away from rivet holes, AA 2024-T3

Table 5.7 Typical pit profile information from Panel #1, coated, AA 2024-T3

No.	Width [μm]	Height [μm]	No.	Width [μm]	Height [μm]
1	38.469	13.38	17	112.063	15.244
2	98.682	10.035	18	85.302	15.746
3	72.757	9.098	19	86.974	16.249
4	146.351	11.239	20	63.558	15.411
5	133.807	18.316	21	54.359	20.772
6	72.757	14.515	22	134.643	17.756
7	99.519	14.86	23	113.736	13.736
8	40.978	9.389	24	217.408	11.341
9	38.469	9.389	25	217.436	21.657
10	26.761	7.861	26	224.962	15.08
11	45.996	7.861	27	45.996	10.909
12	114.572	9.215	28	117.917	16.346
13	68.576	14.525	29	66.067	12.615
14	50.177	9.996	30	168.095	16.168
15	71.085	13.588	31	83.629	20.255
16	66.903	13.276			

5.6.2 2D Profile Information

Pit Profile Information:
In Fig. 5.30, the cross-sectional (CS) area of the highlighted pit profile is 240.43 μm^2.

In Table 5.7, the average pit width is 96.06 μm and the average pit height is 13.74 μm.

In Table 5.8, the average CS area is 354.435 μm^2.

Table 5.8 Typical pit CS areas from Panel #1, uncoated, AA 2024-T3

No.	1	2	3	4	5	6	7	8	9	10
CS area (μm^2)	370.053	430.086	412.393	320.702	384.400	240.430	316.160	340.683	222.708	506.735

5.6.3 3D Profile Information

1. **3D Images**.
2. **3D Surface Roughness Measurement**.

5.7 Cut-off Wavelength λ_c Selection

During an area surface roughness calculation with a profile measurement gauge, irregularities of the surface profile are filtered by introducing an appropriate limiting filter cut-off wavelength λ_c, as indicated in Fig. 5.36. In order to select λ_c, we need to find out defects of interest to us (e.g. pits) size irregularities by analyzing the surface profiles (Figs. 5.31, 5.32 and 5.33).

The surface irregularities in this application are some manual scratches on the panels as shown below. The defects' range is over 700 µm, the profile is as shown in Fig. 5.34, and the defect profile information is compared to a typical pit in the same panel side as shown in Fig. 5.35.

After analyzing a number of corrosion pit profiles and comparing with surface defects (e.g. manual scratches), we chose the cutoff wavelength λ_c of **500 µm** for the area surface roughness calculation.

From Fig. 5.36 it is obvious that when λ_c of 500 µm is applied, the majority of the interest surface features are captured by the filtered surface roughness image, instead of the waviness image.

5.8 Deep Learned Features (DLF)

The widely variable states of corrosion are notoriously difficult to measure and detect, especially at early stages when an insidious problem does not surface until catastrophe strikes. Fast, accurate, automated expert assessment of corrosion can help minimize damage by guiding the condition-based maintenance of structures, as well as the design and use of sensors, from development in the lab to field deployment on aircraft, bridges, etc. A required essential capability in such expert systems is the extraction of features (patterns) from measurements.

Traditionally, feature formulas are hand-engineered by domain experts. For example, in a black-and-white image of pitting formations, a measure of roundness of the contours of pits can be devised. But this assumes that we can automatically delineate the contours of each pit to begin with—useful at later stages of processing but circular logic at the beginning. The issue in this example is that roundness is a feature of the objects of interest (pits), not a feature of a whole image. It is important to be able to examine *both* local features (e.g., to detect potentially costly small cracks) and global features such as texture-based averages (e.g., if condition is

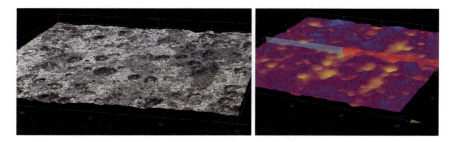

Fig. 5.31 3D profile image from Panel #1, uncoated side, AA 2024-T3

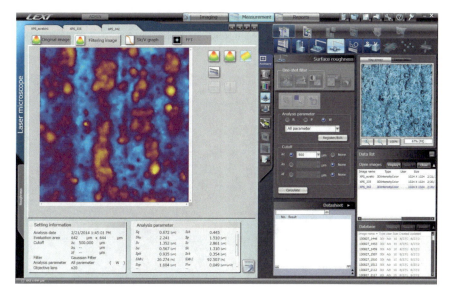

Fig. 5.32 Software interface for area surface roughness calculation

uniform corrosion then situation is relatively benign). An even better approach would be to augment this capability with features that are *hierarchically* represented (covering a spectrum from local to global) and *automatically* learned, as happens in natural biological vision systems [LeCun cat's visual cortex].

Recent breakthroughs in machine learning have enabled unsupervised (i.e., from input data only, without notion of desired outputs) feature learning and massively larger neural networks to be built from training data than was originally possible (e.g., with billions of parameters), in a branch of AI loosely termed Deep Learning [Hinton 2006, DL papers, …]. Our goal is to develop a Deep-Learned Features (DLF) framework that automatically learns neural features from data, to enhance the engineered local/global feature library for corrosion. Advantages of this approach include: (1) ability to learn from data even if a large fraction is *unlabeled*

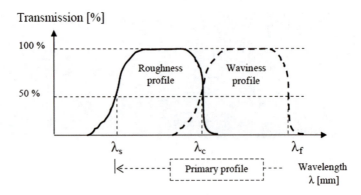

Fig. 5.33 Range of surface profile components with standard separation of waviness and roughness according to filtration of irregularities by cut-off wavelengths

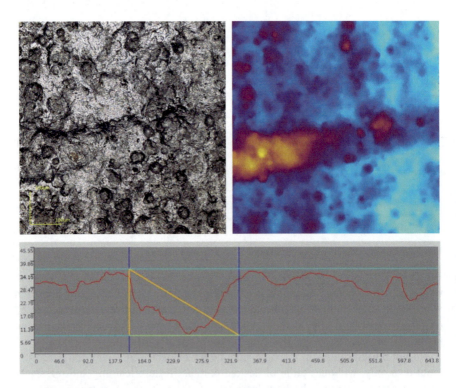

Fig. 5.34 Surface defects in the corroded panel 1 of AA 2024-T3

(no ground-truth labels/classes of corrosion available); (2) ability to use the whole grayscale or color space without necessarily relying on black-and-white binarization (on which morphological filters work best); (3) *scalability* to very large

5 Data Analytics for Corrosion Assessment   147

No.	Result	Width[μm]	Height[μm]	Length[μm]
✓ 1		172.979	27.238	175.111
✓ 2		56.459	15.967	58.674
✓ 3		133.580	19.489	134.995

Fig. 5.35 Profile size comparison (No. 1, 3 corresponding to the manual scratch, and No. 2 corresponding to a typical corrosion pit)

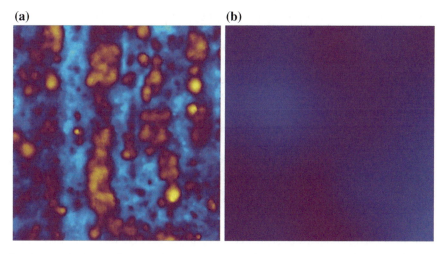

Fig. 5.36 2D height intensity images for area surface roughness calculation when λ_c of 500 μm applied. **a** Surface roughness image. **b** Waviness image

problems (e.g., 1000-class recognition, using GPUs if needed); (4) hierarchical representations in which some neurons may respond selectively to particular localized corrosion problems, as well as more global, protodetector types of features to include in our library. A disadvantage is that the method tends to require many examples (order to thousands) before *generalizing* (i.e., able to also work on unseen data) features can be distilled.

Similar technology is being researched and commercially developed by several companies. For example, forms of deep learning networks were used in Google Brain to automatically learn neurons that respond to cats from YouTube video thumbnails, in Google + Photo Search to detect over 1000 objects of scenes without metadata, in Android speech recognition, in parts of IBM Watson, and in a Microsoft real-time English-to-Mandarin translation demonstration. Facebook, Yahoo, and others have expressed interest at NIPS conferences. Additionally, several image classification public competitions have been won by groups using ensembles of deep networks [refs].

5.9 Methods

Our description is geared towards image data, however, similar principles are applicable to other forms of raw input data, such as time-series measurements obtained from μLPR or other sensors.

A classic solution to classification, regression, or PDF estimation problems using artificial neural networks (NNs) involved the 2- or 3-layer multilayer perceptron architecture, a training dataset of M examples with input vectors $\mathbf{x} \in \mathbb{R}^N$ and desired outputs $\mathbf{y} \in \mathbb{R}^P$, $\{(\mathbf{x}, \mathbf{y})^{(1)}, (\mathbf{x}, \mathbf{y})^{(2)}, \ldots, (\mathbf{x}, \mathbf{y})^{(M)}\}$ (the fact that us teachers present desired outputs is what makes this problem "supervised learning"), trained using backpropagation to simultaneously adjust the weight matrices $\mathbf{W}_i$ and bias vectors $\mathbf{b}_i$ of each layer so as to minimize mean squared error $E\{\mathbf{y} - \hat{\mathbf{y}}^2\}$ between inputs and outputs, possibly regularized to encourage generalization over as yet unseen input data. A two-layered network computes the function $\hat{\mathbf{y}}\sigma(\mathbf{W}_2\sigma(\mathbf{W}_1\mathbf{x} + \mathbf{b}_1) + \mathbf{b}_2)$, where $\sigma(z) = 1/(1 + \exp(-z))$ is the logistic sigmoid function.

We can think of an input image as a long column vector $\mathbf{x}$ containing the pixel intensities (for RGB color, triple the length). Each the K rows of $\mathbf{W}_1$ can be seen as the coefficients of a linear filter with which the dot product between image and filter is computed. The output $o \in [0,1]$ of the sigmoid activation function, after passing thru it the dot product plus bias, can be interpreted in our framework as a *neural feature*. Our DLF framework shares commonalities with the traditional NN setup, but tends to differ in emphasis (harvesting of the neural features inside NN, instead of the NN outputs), supervision levels required for learning (staged between unsupervised and supervised), objective functions being optimized (e.g., addition of sparsity term), use on nontrainable sublayers (e.g., convolution, subsampling, local contrast normalization, etc.), and scalability to more and bigger layers working around the issue of vanishing or exploding gradients that plagued traditional NNs. Our DLF approach includes the following basic strategies:

- Obtain very large number of possibly unlabeled (i.e., without known/ground-truth output) images, and some labeled images (i.e., with desired output classes $\mathbf{y}^{(m)}$, such as $[0\ 0\ 1\ 0]^T$ indicating the 3rd class in a 4-class problem).
- Preprocess input images to have zero mean, and decorrelated and equally scaled dimensions. This is known as zero-phase component analysis (ZCA) whitening, a type of sphering.
- Unsupervised-learning stage:
  1. From unlabeled data, learn neural features via sparse autoencoder (AE) using backpropagation.
  2. If necessary, greedily train more AEs in a stack (i.e., pretrain each AE independently of rest of the network, as opposed to all layers jointly).
- Supervised-learning stage: From labeled data, train a classification layer and/or tune whole network using backpropagation.
- Scale up to big images (probably anything above 64 × 64) by randomly sampling small patches (e.g., 16 × 16) and inserting convolution and pooling operations between layers.

We provide more details of these strategies next.

Prewhitening
Sparse Autoencoder

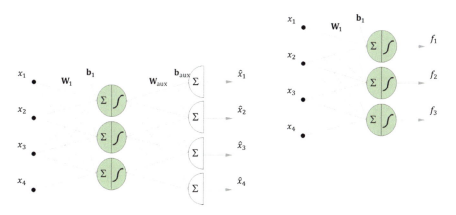

Minimize the scalar objective function

$$J = \frac{1}{2}\left\langle \|\mathbf{x} - \hat{\mathbf{x}}\|^2 \right\rangle + \frac{\lambda}{2}\sum\left(w_{ij}^{(l)}\right)^2 + \beta\sum KL(\rho\|\hat{\rho}_j)$$

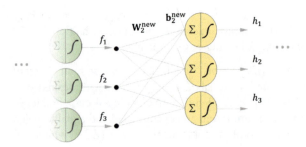

Convolutional Feature Extraction

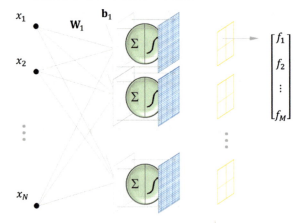

Supervised Stage

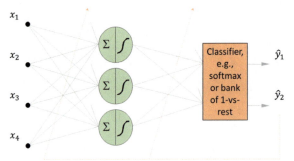

5.10 Codebase Validation

Starting from the stanford_dl_ex MATLAB stubs [Ng refs], we derived all the equations and generated and unit-tested all the code for sparsity-based MSE and cross-entropy cost functions, efficient backpropagation-based gradients, stacked AEs, convolutional and pooling layers, mini-batch stochastic gradient descent, required for the DLF framework. We validated the codebase with several experiments using known real-world big data sets including CIFAR-10, MNIST, and STL-10, achieving expected accuracies from ~ 81 to 98% when tested over thousands of independent examples unseen during training. We have tens of images of corrosion states which is not yet enough to extract generalizing features. Until thousands of images can be obtained, we also verified the applicability of the method to corrosion using simulated sticks versus circles as proxies for cracks and pits.

Synthetic Images Verification for Corrosion—the following figures illustrate the procedural steps for corrosion image verification.

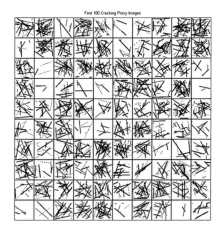
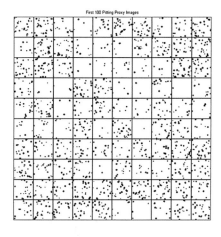

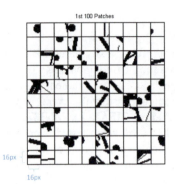

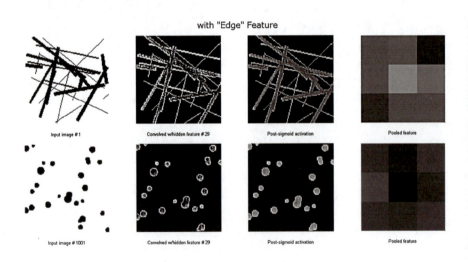

5.11 Conclusion

Extending this work to thousands of corrosion images (from confocal microscopy in the lab plus photorealistic simulation), our DLF framework is poised to add powerful, deep-learned feature vectors (e.g., the 9-dimensional pooled features in figure above) to the engineered feature library. Future investigations should examine other hybrid architectures in which, for example, DLF serves as 1st-pass processor to guide the segmentation of regions of interest in corrosion images, a pre-requisite for the engineered local features to work well.

5.12 Feature Selection

Feature selection methods fall under three main categories: wrapper methods, embedded methods, and filter methods.

Wrapper methods assign value to as set of features by the performance of the data mining algorithm. Figure 5.37 shows a schematic of a generic wrapper method. The value of the subset of extracted features in this case is the accuracy of the data mining task using the subset of features from the training set. D. Garrett et al. implemented this approach for a classification task by using a genetic algorithm to search through the extracted feature space and a support vector machine to perform the classification [4]. Other popular search strategies in the literature include: best-first, branch-and-bound, and simulated annealing [5]. In addition, decision trees, naïve Bayes, and least-square linear predictors are popular data mining algorithms used for performance evaluation [5]. Table 5.5 lists various data mining algorithms that can be used to assess the performance of a set of features.

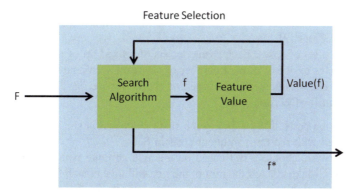

Where F is the set of extracted features, f ⊂ F, and f* is the subset of F with the highest value

Fig. 5.37 Schematic of general wrapper feature selection method

One of the main disadvantages of wrapper methods is that they can be computationally intensive.

In embedded methods the search for the optimal subset of extracted features is built into the classifier construction. Therefore, the search is in the combined space of feature subsets and hypotheses for the classification.

Filter methods assign value to features only by looking at intrinsic properties of the data and are independent of the chosen data mining algorithm. Feature ranking is a filter method for feature selection. Consider a supervised learning problem where there is a set of m observations $\{f_{1,k},\ldots,f_{n,k},y_k\}$ (k = 1,..., m). Where $\{f_1,\ldots,f_n\}$ is the set of n extracted features and y is the desired output. The value of f_i is computed using a scoring function which compares f_i to y. For example, for linear regression the squared value of the estimate of the Pearson Correlation Coefficient is used as the scoring function. The estimate of the Pearson Correlation Coefficient is given as:

$$R(i) = \frac{\sum_{k=1}^{m}(f_{i,k}-\bar{f}_i)(y_k-\bar{y})}{\sqrt{\sum_{k=1}^{m}(f_{i,k}-\bar{f}_i)^2 \sum_{k=1}^{m}(y_k-\bar{y})^2}}$$

where the bar notation stands for an average over the index k. $R(i)^2$ is used as a scoring function since it represents the fraction of total variance around the mean value $\bar{y}$ that is explained by the linear relation between f_i and y [5]. Table 5.9 lists several data mining methods. Table 5.10 lists advantages, disadvantages and examples as pointed out by Y. Saeys et al. in their review of feature selection techniques.

5.13 Classification Techniques

- Decision Trees [2]
  Let S be a set of cases. Each case consists of a fixed set of attributes (features) and belongs to one of a small number of classes. Given a new case the decision tree will classify the new case based on the known set S.
  Algorithm for Decision Tree:
  - If all the cases in S belong to the same class or S is small, the tree is a leaf labeled with the most frequent class in S.
  - Otherwise, choose a test based on a single attribute with two or more outcomes. Make this test the root of the tree with one branch for each outcome if the test, partition S into corresponding subsets $S_1, S_2,\ldots$ according to the outcome for each case, and apply the same procedure recursively to each subset.

5 Data Analytics for Corrosion Assessment

Table 5.9 Data mining methods

Method	Description	Techniques/algorithms
Classification [1, 2]	Learn a function that maps (classifies) a data item into one of several predefined classes	• Decision trees • Rule set classifiers • Support vector machines (SVM) • Nearest neighbor • Case based reasoning • Neural networks • AdaBoost • Naive Bayes

Regression [1]	Learn a function that maps a data item to a real-valued prediction variable	• Linear regression
		  – Least squares
		• Non-linear regression
		  – Feed-forward neural networks
		  – Adaptive spline methods
		  – Projection pursuit regression

Clustering [1, 2]	Identify a finite set of categories or clusters to describe the data	• K Means • Expectation maximization algorithm (EM)
Summarization [1]	Involves methods for finding a compact description for a subset of data. For example tabulating the mean and standard deviation for all fields	• Calculated mean and standard deviation of all fields • Multivariate visualization techniques
Dependency modeling [1]	Find a model that describes the significant dependencies between variables	• Probabilistic dependency networks

- Ruleset classifiers [2]
  - Consists of a list of rules of the form "if A and B and C ... then class X", where rules for each class are grouped together. A case is classified by finding the first rule whose conditions are satisfied by the case; if no rule is satisfied, the case is assigned to a default class.
  - Rulesets are formed from the initial (unpruned) decision tree. Each path from the root of three to a leaf becomes a prototype rule whose conditions are the outcomes along the path and whose class is the label of the leaf. This rule is then simplified by determining the effect of discarding each condition in turn.
  - A subset of simplified rules is selected for each class in turn.
  - Principal disadvantage is the amount of CPU time and memory required.

Table 5.10 Advantages, disadvantages and examples of feature selection methods [6]

Method		Advantages	Disadvantages	Examples
Filter	Univariate			
		• Fast • Scalable • Independent of classifier	• Ignores feature dependencies • Ignores interaction with the classifier	• χ^2 • Euclidean distance • I-test • Information gain, Gain ratio
	Multivariate			
		• Models feature dependencies • Independent of the classifier • Better computational complexity than wrapper methods	• Slower than univariate techniques • Less scalable than univariate techniques • Ignores interaction with the classifier	• Correlation-based feature selection • Markov blanket filter • Fast correlation-based feature selection
Wrapper	Deterministic			
		• Simple • Interacts with the classifier • Models feature dependencies • Less computationally intensive than randomized methods	• Risk of over fitting • More prone than randomized algorithms to getting stuck in a local optimum (greedy search) • Classifier dependent selection	• Sequential forward selection (SFS) • Sequential backward elimination (SBE) • Plus q take-away r • Beam search
	Randomized			
		• Less prone to local optima • Interacts with the classifier • Models feature dependencies	• Computationally intensive • Classifier dependent selection • Higher risk of overfitting than deterministic algorithms	• Simulated annealing • Randomized hill climbing • Genetic algorithms • Estimation of distribution algorithms
Embedded		• Interacts with the classifier • Better computational complexity than wrapper methods • Models feature dependencies	• Classifier dependent selection	• Decision trees • Weighted Naïve Bayes • Feature selection using the weight vector of SVM

- Support vector machines (SVM) [2]

    - In a two-class learning task, the aim of SVM is to find the best classification function to distinguish between members of the two classes in the training data.
    - For a linearly separable dataset, a linear classification function corresponds to a separating hyper-plane f(x) that passes through the middle of the two classes, separating the two.

- SVM finds the best function that maximizes the margin between the two classes.
- An SVM classifier attempts to maximize the following function with respect to $\vec{w}$ and b: $L_p = \frac{1}{2}\left\|\vec{w}\right\| - \sum_{i=1}^{t} \alpha_i y_i \left(\vec{w} \bullet \vec{x}_i + b\right) + \sum_{i=1}^{t} \alpha_i$

where t is the number of training samples, and α_i, i = 1,...,t are non-negative numbers such that the derivatives of L_p with respect to α_i are zero.

Clustering Techniques:

- K-means algorithm [2]
  The algorithm operated on a set of d-dimensional vectors, $D = \{x_i | i = 1, \ldots, N\}$, where $x_i \in \mathbb{R}^d$ denotes the ith data point.
- The algorithm is initialized by picking k points in $\mathbb{R}^d$ as the initial k cluster representatives or "centroids".
- The algorithm then iterated between two steps until convergence:
  - Step 1: Data Assignment. Each data point is assigned to its closest centroid, with ties broken arbitrarily. This results in a partitioning of the data.
  - Step 2: Relocation of the "means". Each cluster representative is relocated to the center (mean) of all the data points assigned to it.
- Disadvantages of the algorithm:
  - Sensitive to initial centroids
  - Fill falter when the data if not well described by reasonably separated spherical balls

  Sensitive to outliers.

5.14 Sensor Data Fusion

Although significant achievements have been reported in the recent past, the processing of sensor data intelligently still requires the development, testing, and validation of new techniques to manage and interpret the increasing volume of data and to combine them as they become available from multiple and diverse sources. Sensor data fusion is a promising technology that can contribute significantly towards a better understanding and a more efficient utility of raw data by reducing it to useful information [17]. We are introducing in this proposal new and innovative fusion techniques that build upon current data management practices but also advance the state of the art through a systems engineering process that is rigorous and verifiable. We define the fusion problem in a generic framework. A methodology is sought using intelligent decision making tools through which data collected from a variety of sensors under various testing, modeling or field conditions can be aggregated in a meaningful and systematic way to provide information to the decision makers at the operational task level. We synthesize the information to higher informational levels. A typical sensor data fusion paradigm incorporates several levels of abstraction: fusion at the data level, the feature (characteristic signature of the fault or failure data) level, the sensor level and the knowledge level. At the data level, a variety of filtering, data compression and data validation algorithms are employed to improve such indicators as signal to noise ratio, among others. The enabling technologies at the feature level borrow from Dempster-Shafer theory, soft computing and Bayesian estimation to fuse feature while meeting specified performance metrics [18]. At the sensor level, we rely upon concepts from information theory while multiple sensors are gated and coordinated spatially and temporally to minimize their number while maximizing the probability of detection. Significant reduction of the computational burden is always a desired objective. The top level of the fusion hierarchy, i.e. the knowledge fusion module is designed to reason about the evidence provided by the lower echelons, aggregate the available information in an intelligent manner, resolve conflicts and report to the end-use the findings of the fusion architecture. Artificial Intelligences (AI) tools and methods from Dempster-Shafer theory, Bayesian estimation techniques and soft computing may find utility as the reasoning enablers at this level.

5.14.1 Fusion at the Feature Level

Feature fusion has attracted the attention of the research community in recent years, as data and data acquisition/processing strategies proliferate, in order to maximize the value of information extracted from raw data while improving the algorithms' computational efficiency [19–22]. Fused features (meta-features, synthetic features, as sometimes called) improve the performance of diagnostic and prognostic algorithms by increasing the correlation of the constituent features with respect to ground truth fault data. The feature level offers the most advantageous and beneficial opportunity for the application of novel fusion techniques. It is common practice to employ high-bandwidth dynamic sensing modalities (vibration, dynamic pressure, etc.) in order to monitor key attributes of fault/failure modes for critical components/subsystems.

The fundamental principle of fusion algorithms is rather simple: maximize a utility or objective/fitness function that conveys the relationship between the fused features and the actual fault dimension. A number of fusion algorithms, fitness functions and optimization solvers have been proposed over the past years. The challenge stems primarily from the need to define an optimum feature vector and to select the most appropriate fitness function for the problem at hand. The fitness function generally attempts to capture the "similarity" between the fused features and ground truth.

Typical features or CIs in the time domain may include peak values, RMS, energy, kurtosis, etc. in the frequency domain, we focus primarily on features for rotating equipment that exhibit a marked difference between baseline or no-fault and faulty data [23]. For example, we seek in this category a comparison (amplitude, energy, etc.) of certain sidebands to dominant frequencies, when the sensor signals are transformed via an FFT routine to the frequency domain[20]. Other possible features are extracted through coherence and correlation calculations. When the information is shared between the time and frequency domain, it might be advantageous to extract features in the wavelet domain offering an appropriate tradeoff between the two domains. When multiple features are extracted for a particular fault mode, it is desirable to combine or fuse uncorrelated features to enhance the fault detectability.

Consider the objective or fitness function:

$$f = |correlation(x, w)| \times FDR(x, w)$$

where x and w could represent two features.

Other suitable fitness functions may be defined depending on the problem at hand. The choice of the "best" fitness function is a challenging task and the most significant step in the optimization process. Once the fitness function and appropriate initial conditions in the search space are given, the algorithm (PSO, GP, etc.) is allowed to run until specified termination conditions are satisfied.

Well established tools like Principal Component Analysis (PCA) and methods based on the degree of overlap between the probability density functions of features are employed first to screen features, prioritize and rank them for further processing. Thus, we face eventually only a subset of the "best" features extracted from raw data.

It is preferable to view features and their corresponding fused versions in a probabilistic or statistical setting. Bayesian estimation methods, Kalman filtering, particle filtering, etc., allow for information from multiple measurement sources to be fused in a principled manner. Typically, multiple snapshots or windowed data are used to extract a feature. A histogram is built next from the feature sequence approximating a Probability Density Function (PDF). Similar constructs are determined from other features from baseline and fault data. Since fusion, regardless of the method employed, is viewed as an optimization problem, an appropriate fitness or objective function must be defined to evaluate each feature. In our brain research work, we defined and employed successfully the following fitness function [24]:

$$fitness = \frac{\sqrt{\sigma_1^2 + \sigma_2^2}}{|\mu_1 - \mu_2|} \times \left(\frac{1}{1 - PDF_{overlap}} \right)$$

where μ_1, μ_2 are the means and σ_1, σ_2 the standard deviations of two features. The PDF overlap is the common area between the given feature PDF and one obtained from data under no fault conditions. Here, we are attempting to discriminate or distinguish between features belonging to the two classes only: One class representing baseline or no fault conditions and the second representing a faulty stat. Thus, the fitness function is composed of the inverse square root of the Fisher Discriminant Ratio (FDR) divided by one minus the PDF overlap [25].

FDR measures the distance between two classes of features. Sometimes seemingly good FDR values still produce features with large overlaps in the feature histograms for the two classes. This fitness function penalizes features with large class overlaps by increasing the fitness score in proportion to the amount of overlap. The overlap values range from 0 for no overlap to 1 for total overlap. The feature with the lowest fitness score is selected as the best.

5.15 Epilogue

This chapter describes rigorous tools/methods for processing corrosion imaging data. An array of approaches to data mining, feature extraction and selection and classification are presented with extensive examples illustrating the efficacy of these methods. The enabling technologies cover a wide spectrum of innovative legacy and new advances like Deep Learned Features borrowed from the Deep Learning domain. The contents of this chapter set the stage for accurate corrosion detection and prediction algorithms.

References

1. Hoeppner DW, Chandrasekaran V, Taylor AMH (1999) Review of pitting corrosion fatigue models. In: International committee on aeronautical fatigue, Bellevue, WA, USA
2. Kawai S, Kasai K (1985) Considerations of allowable stress of corrosion fatigue (focused on the influence of pitting). Fatigue Fract Eng Mater Struct 8(2):115–127
3. Lindley TC, Mcintyre P, Trant PJ (1982) Fatigue-crack initiation at corrosion pits. Metals Technol 9(1):135–142
4. Pidaparti RM (2007) Strucural corrosion health assessment using computational intelligence methods. Struct Health Monit 6(3):245–259
5. Frankel GS (1998) Pitting corrosion of metals: a review of the critical factors. J Electrochem Soc 145(6):2186–2198
6. Huang T-S, Frankel GS (2006) Influence of grain structure on anisotropic localized corrosion kinetics of AA7xxx-T6 alloys. Corros Eng Sci Technol 41(3):192–199
7. Szklarska-Smialowska Z (1999) Pitting corrosion of aluminum. Cossorion Sci 41(9): 1743–1767
8. Pereira MC, Silva JW, Acciari HA, Codaro EN, Hein LR (2012) Morphology characterization and kinetics evaluation of pitting corrosion of commercially pure aluminum by digital image analysis. Mater Sci Appl 3(5):287–293
9. Clark T (2009) Guided wave health monitoring of complex structures. Imperial College London, London
10. Belanger P, Cawley P, Simonetti F (2010) Guided wave diffraction tomography within the born approximation. IEEE Trans UFFC 57:1405–1418

11. Li H, Michaels JE, Lee SJ, Michaels TE, Thompson DO, Chimenti DE (2012) Quantification of surface wetting in plate-like structures via guided waves. AIP Conf Proc Amn Inst Phys 1430(1):217
12. Forsyth DS, Komorwoski JP (2000) The role of data fusion in NDE for aging aircraft. In: SPIE aging aircraft, airports and aerospace hardware IV, vol 3994, p 6
13. Brown D, Darr D, Morse J, Laskowski B (2012) Real-time corrosion monitoring of aircraft structures with prognostic applications. In: Annual conference of the prognostics and health management society, vol 3
14. Brown DW, Connolly RJ, Laskowski B, Garvan M, Li H, Agarwala VS, Vachtsevanos G (2014) A novel linear polarization resistance corrosion sensing methodology for aircraft structure. In: Annual conference of the prognostics and health management society, vol 5, no 33
15. McAdam G, Newman PJ, McKenzie I, Davis C, Hinton BR (2005) Fiber optic sensors for detection of corrosion within aircraft. Struct Health Monit 4:47–56
16. Sharland SM (1987) A review of the theoretical modeling of crevice and pitting corrosion. Pergamon J Ltd 27(3):289–323
17. Liggins M, Hall DL, Linas J (2008) Multisensor data fusion: theory and practice. CRC Press, Boca Raton
18. Dar I, Vachtsevanos G (1997) Feature level sensor fusion for pattern recognition using an active perception approach. In: Proceedings of IS&T/SPIE's electronic imaging '97: Science and Technology, San Jose, CA, February 8–14, 1997
19. Vachtsevanos G, Lewis G, Roemer M, Hess A (2006) Intelligent fault diagnosis and prognosis for engineering systems. Wiley, Hoboken
20. Wang F, Sun F, Cao B (2007) Feature fusion of mechanical faults based on evolutionary computation. Insight 49(8):471–475
21. Voulgaris Z, Sconyers C, Vachtsevanos G (2006) A particle swarm optimization approach to feature fusion for failure prognosis of engineering systems. In: Proceedings of 18th mediterranean conference on control and automation, June 2010, Morocco, pp 339–344
22. Wu B, Saxena A, Partrick R, Vachtsevanos G (2005) Vibration monitoring for fault diagnosis of helicopter planetary gears. In Proceedings of the 16th IFAC
23. López De La Cruz J, Lindelauf RHA, Koene L, Gutiérrez MA (2007) Stochastic approach to the spatial analysis of pitting corrosion and pit interaction. Electrochem Commun 9(2): 325–330
24. Burrell L, Litt B, Vachtsevanos G (2007) Evaluation of feature selection techniques for analysis of functional MRI. In Proceedings of the 2007 international conference on data mining, Las Vegas, NV, June 25–28, 2007
25. Burrell L, Glynn S, Vachtsevanos G, Litt B (2007) Feature analysis of functional MRI for discrimination between normal and epileptogenic brain. In: European control conference, Kos, Greece

Chapter 6
Corrosion Modeling

George Vachtsevanos

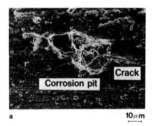 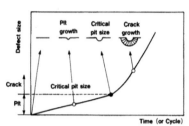

Abstract This chapter introduces fundamental concepts, methods and techniques for modeling of corrosion processes. Modeling and simulation platforms constitute the cornerstone for corrosion prevention, remediation and diagnosis/prognosis of corrosion initiation and propagation in critical metal structures. We address methods that cover first principle models, semi-empirical and empirical models for corrosion processes. We begin with the electrochemical nature of corrosion as the basis for microscale modeling efforts and proceed to describe mesoscale and macroscale modeling techniques. The absence of real long-time corrosion data derived from field testing, laboratory experiments and high-confidence models (mostly absent in the current literature) necessitates the development and application of mostly empirical approaches. The modeling approaches reported in the literature are reviewed and summarized.

Keywords Corrosion imaging · Corrosion processes · Micro-scale corrosion models · Meso-scale corrosion models · Macro-scale corrosion models · Data-driven corrosion models · Model-based approaches · Stochastic corrosion models · Corrosion failure prediction models · Global versus local corrosion models

---

G. Vachtsevanos (✉)
Georgia Tech, Atlanta, USA
e-mail: gjv@ece.gatech.edu

6.1 The Need

Reliable, high-fidelity corrosion models form the foundation for accurate and robust corrosion detection and growth prediction. A suitable modeling framework assists in the development, testing and evaluation of detection and prediction algorithms. It may be employed to generate data for data-driven methods to diagnostics/prognostics, test and validate routines for data processing tool development, among others. The flexibility provided by a simulation platform, housing appropriate detection and progression models, is a unique attribute in the study of how corrosion processes are initiated, evolving and may be, eventually, mitigated in physical systems. There is evidence to support that aircraft, ships, transportation systems and industrial processes are subjected to severe corrosion that is costing billions of dollars to prevent and/or remedy.

The pictures below show aircraft failures attributed to corrosion.

(media photos)

6.2 The Objective

The objective of the corrosion modeling effort is to assess the current state of the art and develop, test and evaluate novel corrosion detection and progression models that will assist in the design and implementation of "smart" sensors and sensing modalities for a variety of application domains. We will capitalize on novel modeling activities to assure that we arrive at an intelligent sensing framework for corrosion detection and growth. Our ultimate objective is to assist in development of an integrated "smart" sensing strategy that will be capable of detecting accurately

6 Corrosion Modeling

corrosion initiation and will be capable of predicting on-line the corrosion growth rate. On the imaging side, in addition to corrosion modeling based on classical EO images, we will be exploring other means for assembling "images" via a network of multiple sensors. Bring-to-market a corrosion monitoring system that is,

- Capable of detection and prediction of pit/ crevice initiation.
- Suitable for installation on multiple platforms: low power, long lifetime, small size, low weight, wireless, low EMI susceptibility.

The figure below depicts basic corrosion processes (Fig. 6.1; Table 6.1).

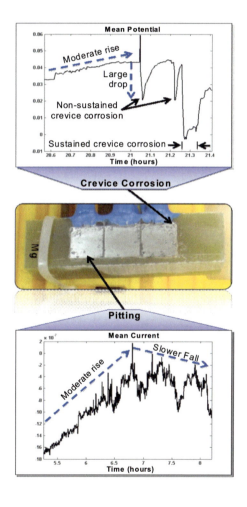

Fig. 6.1 Basic corrosion process

Table 6.1 Pitting corrosion incidents of aircraft and helicopters [1]

Aircraft	Location of failure	Cause	Incident severity	Place	Year	From
Bell helicopter	Fuselage, longeron	Fatigue, corrosion and pitting present	Serious	AR, USA	1997	NTSB
DC-6	Engine, master connecting rod	Corrosion pitting	Fatal	AK, USA	1996	NTSB
Piper PA-23	Engine, cylinder	Corrosion pitting	Fatal	AL, USA	1996	NTSB
Boeing 75	Rudder control	Corrosion pitting	Substantial damage to plane	WI, USA	1996	NTSB
Embraer 120	Propeller blade	Corrosion pitting	Fatal and serious, loss of plane	GA, USA	1995	NTSB
Gulfstream GA-681	Hydraulic line	Corrosion pitting	Loss of plane, no injuries	AZ, USA	1994	NTSB
L-1011	Engine, compressor assembly disk	Corrosion pitting	Loss of plane, no injuries	AK, USA	1994	NTSB
Embraer 120	Propeller blade	Corrosion pitting	Damage to plane, no injuries	Canada	1994	NTSB
Embraer 120	Propeller blade	Corrosion pitting	Damage to plane, no injuries	Brazil	1994	NTSB
F/A-18	Trailing-edge flap (TEF) outboard hinge lug	Corrosion pitting, fatigue	Loss of TEF	Australia	1993	AMRL
Mooney Mooney 20	Engine, interior	Corrosion pitting, improper approach	Minor injuries	TX, USA	1993	NTSB
Aero Commander 680	Lower spar cap	Corrosion pitting	Fatal	Sweden	1990	Swedish CAA

6.3 Corrosion Books

A number of books addressing topics from corrosion principles to corrosion science, corrosion data and reference treatises have been published over the past years. Most of them address specific topics in corrosion science and engineering while others focus on corrosion data or target a more general audience as reference books.

Corrosion Monitoring is addressed in another chapter. A typical setup for corrosion monitoring on an aircraft structure is shown below.

6.4 The Data Base

A suitable and statistically sufficient database of corroding devices/structures/panels constitutes the necessary foundation for corrosion modeling. Data derived from testing of corroding specimens or field data, if available, and imaging tools form the foundational basis for corrosion modeling. We list below a sample of corrosion data assembled from a variety of sources.

- Factors:
  - Temperature.
  - Relative humidity.
- pH is assumed to be known a priori:
  - pH: 7—Distilled water.
  - pH: 7—Salt water (5% NaCl).
- By definition, time of wetness is the amount of time an environment exceeds 80% of Relative Humidity (RH).

- Corrosion rate (CR) varies directly with temperature and RH and indirectly with pH (more acetic yields more corrosion).
- Corrosion is measured in mass loss per surface area (grams/cm^2).
- Pit depth is computed from mass loss surface density and geometry of the anticipated pit.

6.5 Imaging Data

It is evident that imaging of corroding surfaces offers a viable, robust and accurate means to assess the extent of localized corrosion [2]. Modeling, in combination with sensor measurements, promises to assist in the timely and accurate corrosion prevention, detection and prediction.

Corrosion fatigue is recognized as one of the degradation mechanisms that affect the structural integrity of aging aircraft structures [3]. Several nondestructive inspection (NDI) systems (eddy current, ultrasound and others) have been used to obtain the images of damaged regions. There is a growing demand for improving existing NDI techniques to achieve maximum confidence and reliable results with minimum damage components. There is always a constant outlook for methods that identify the damaged regions on the image and also that gives a quick estimate of the extent of the damage.

The images obtained through conventional NDI methods are not directly suitable for identification and quantification of damaged regions. Such images therefore need to be enhanced and segmented appropriately for further image analysis. Segmentation has been achieved using wavelet decomposition (see chapter on Data Analytics). The wavelet coefficients also can be used to quantify the extent of corrosion. Neural networks were applied in the process of segmentation and quantification of damaged regions. Segmentation results show a good correspondence between the extracted regions and the actual damaged on sample panels. We show below (Figs. 6.2, 6.3 and 6.4) samples of structural specimens imaged via conventional tools. Simple image processing tools/results are also included in Fig. 6.4.

Fig. 6.2 Samples of structural specimens. *Source* Danny Parker AVNIK/US Army

Fig. 6.3 Samples of structural specimens. *Source* Danny Parker AVNIK/US Army

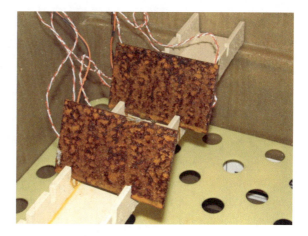

Fig. 6.4 Cropped image

6.6 Salt Fog Images

This section illustrates initial image processing/analysis methods/results. Figures 6.5, 6.6 and 6.7 show material from an aerospace wing attachment fitting that was subjected to 69 h of salt fog. In order to enhance the contrast of corrosion images a histogram equalization was applied to the images. The results of histogram equalization are shown in Fig. 6.5. In order to distinguish areas on the panel of different levels of corrosion thresholds were applied to the corrosion images. The results of applying thresholds to the images are shown in Figs. 6.5, 6.6, 6.7 and 6.8.

Figures 6.9, 6.10 and 6.11 illustrate initial image processing/analysis methods/results on bearing images.

The threshold techniques discussed so far characterize the level of corrosion based primarily on the pixel intensity of the image. However, since the level of

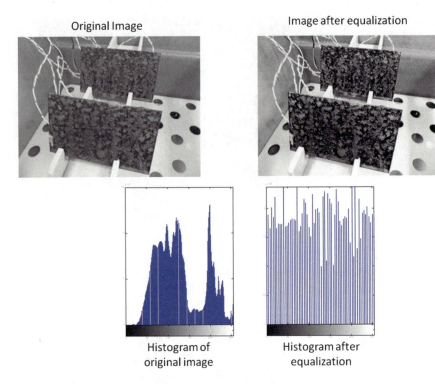

Fig. 6.5 Histogram equalization

Fig. 6.6 Image thresholding

6 Corrosion Modeling

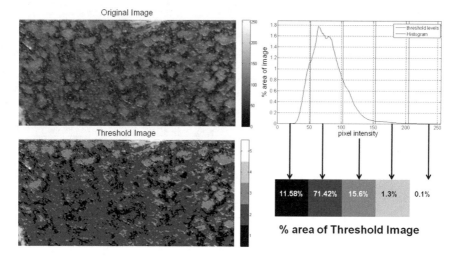

Fig. 6.7 Uniform thresholding of corrosion image

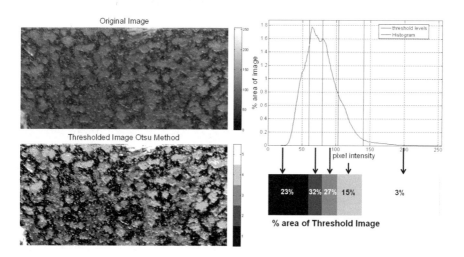

Fig. 6.8 Otsu method for thresholding of corrosion image

corrosion will depend on the texture and not just the pixel value of the image, it is important to consider many features to determine the level of corrosion in different regions of the image. Such features could include texture analysis, two dimensional Fourier transform coefficients, discrete cosine transform coefficients, homogeneity,

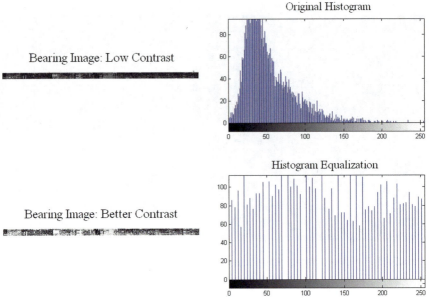

Fig. 6.9 Histogram equalization of corroding bearing image

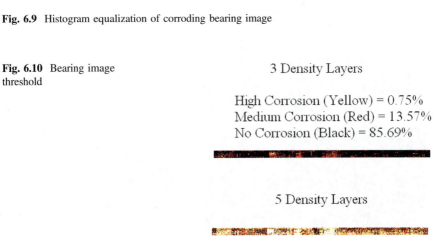

Fig. 6.10 Bearing image threshold

entropy, contrast, range filtering etc. Figure 6.12 demonstrates the usefulness of using two-dimensional Fourier transform coefficients to distinguish between a non-corroded image (baseline) and a corrosion image. Note that the zero frequency components are in the center of the images.

6 Corrosion Modeling 173

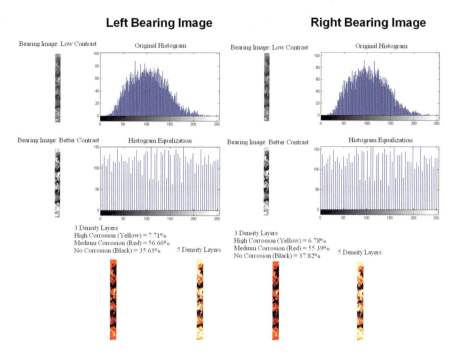

Fig. 6.11 Histogram equalization and threshold on left and right bearing images

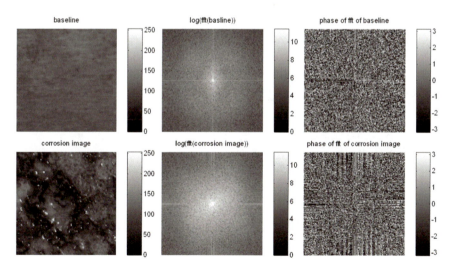

Fig. 6.12 Two—dimensional FFT of baseline and corrosion images

6.7 Fundamental Corrosion Processes

Thermodynamic principles for corrosion attempt to explain why corrosion occurs. Aluminum alloy chemistry and the formation of Grain Boundaries (GBs) in such alloys address fundamental issues/concepts of how alloying elements (micro-constituents) add to strengthening and corrosion resistance. GBs contribute initially to structure defects and the formation of corrosion products at the micro-scale level. Chemical reactions at the GB contribute to pit growth.

The anisotropic grain structure in an alloy-short transverse grain topology at the grain boundaries leads to significant susceptibility to intergranular corrosion and constitutes the initial phase of corrosion initiation and growth.

Initial pitting occurs at this stage that may be detectable via mass loss measurements or imaging techniques [1]. The pit depth at this initial stage is minimal. For modeling purposes, we may assign the designation "corrosion" in order to differentiate this phase from actual/measurable pitting and cracking which significantly affect the state of corrosion and may lead to severe material loss/deformation requiring maintenance action. Thus, a continuous active corrosion path exists at the grain boundaries. Figure 6.13 illustrates the typical corrosion cycle while Fig. 6.14 shows a pictorial representation of grain boundaries. Figure 6.15 depicts pitting corrosion images.

Crevice corrosion (initiation and sequence) is addressed and included in the modeling effort when testing and data become available for alloy structures that exhibit the presence of fasteners, joints, bolts, etc. Moreover, stress corrosion cracking mechanisms (vibration-induced or via other means) will be considered when test panels and lap joints provide appropriate data.

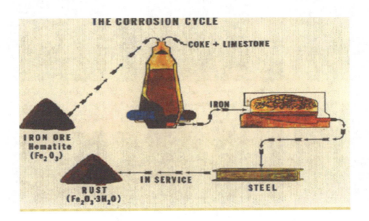

Fig. 6.13 Schematic of the corrosion cycle (*Source* Dr. Vinod Agarwala)

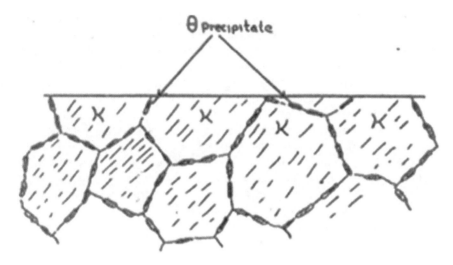

Fig. 6.14 Idealized pictorial of grain boundary precipitation in an aluminum-copper alloy (*Source* Dr. Vinod Agarwala)

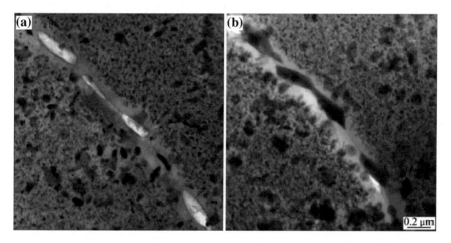

Fig. 6.15 Pitting images (*Source* Dr. Vinod Agarwala)

6.8 Corrosion Modeling: Background/State of the Art

Corrosion modeling has been an active area of research over the recent past. A variety of models have been proposed depending on the material processes, the application domain, etc. We review briefly major advances in this area. The research literature in this area is extensive and our intent is to summarize some important contributions only.

Corrosion modeling has been addressed via a number of methods ranging from empirical, semi-empirical and firs principle approaches. Most efforts focused on empirical techniques with first principle models exploring the electro-chemical nature of corrosion processes.

The US Air Force has undertaken a massive role on aircraft corrosion processes [4]. Of particular interest to this document is the corrosion modeling section detailed in this lengthy paper. The section summarizes the results of the modeling effort to date. An Eddy current technique is commonly used as an NDI (Non Destructive Inspection) for detection of hidden cracks and corrosion in commercial aircraft structures. This technique detects variations in the specimen's ability to generate eddy current in the presence of the time varying magnetic field.

The model integrates the results of environmental modeling, susceptibility modeling, condition modeling, corrosion growth rate modeling, and structural effects modeling. All have been completed and integrated. A combining model has been developed that results in a corrosion prediction that permits determination of present corrosion damage and predicts the effects of future corrosion damage.

An economic effect model has been integrated into the structural model permitting the examination of cost effectiveness of various maintenance actions. Figure 6.16 suggests a basic corrosion mechanism.

It has been shown that corrosion is electrochemical in nature with Fig. 6.16 showing the basic reactions. As a result, research activity in the electro-chemistry community has produced a multiplicity of modeling tools/methods to describe corrosion processes. Our interest is on those practical issues of corrosion modeling that affect the initiation, propagation and prediction of corrosion in engineering systems/processes. It is well established that corrosion is electrochemical in nature. The anodic and cathodic reactions are shown in Fig. 6.17.

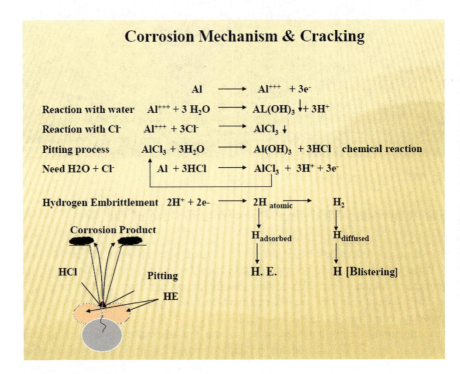

Fig. 6.16 Basic corrosion mechanisms (*Source* Dr. Vinod Agarwala)

> **CORROSION IS ELECTROCHEMICAL IN NATURE**
> General anodic reaction: $M_{(s)} \rightarrow M^{n+} + ne^-$
> **General cathodic reactions:**
> (acid solutions) $O_2 + 4H^+ + 4e^- \rightarrow 2H_2O$
> Hydrogen evolution: $2H^+ + 2e^- \rightarrow H_2{\uparrow}_{(g)}$
> (neutral or basic solutions) $O_2 + 2H_2O - 4e^- \rightarrow 4OH^-$

Fig. 6.17 Fundamental anodic and cathodic reactions (*Source* Dr. Vinod Agarwala)

6.9 Introduction to the Modeling Framework

We exploit novel image processing tools/methods to identify features of interest to be used in the modeling task, since imaging of corroding surfaces offers a viable, robust and accurate means to assess the extent of global or localized corrosion.

The corrosion models address:

- pitting and intergranular corrosion (microscale) of Al alloys and other metals.
- crevice corrosion in occluded areas, such as joints (mesoscale).
- galvanic corrosion of aircraft structural elements (macroscale).
- the effect of surface protection methods (anodization, corrosion inhibitor release, clad layer, etc.).

A systematic, thorough and robust corrosion modeling effort, addressing all corrosion stages for steel, aluminum alloys or other metals, from micro to meso and macro levels, combined with appropriate sensing, data mining and decision support tools/methods (diagnostic and prognostic algorithms) may lead to substantially improved structural component (materials, coatings, etc.) performance and reduced exposure to detrimental consequences [5]. Reliable, high-fidelity corrosion models form the foundation for accurate and robust corrosion detection and growth prediction. A suitable modeling framework assists in the development, testing and evaluation of detection and prediction algorithms. It may be employed to generate data for data-driven methods to diagnostics/prognostics, test and validate routines for data processing tool development, among others. The flexibility provided by a simulation platform, housing appropriate detection and progression models, is a unique attribute in the study of how corrosion processes are initiated, evolving and may be, eventually, mitigated in physical systems.

6.10 Basic Modules of the Smart Sensing Modality and Corrosion Modeling

Generally speaking, corrosion starts in the form of pitting, owing to surface chemical or physical heterogeneity, and then facilitated by the interaction of the corrosive environment fatigue cracks initiate from corrosion pitted areas and further grow into the scale that would lead to accelerated structure failure [6, 7]. In order to effectively conduct structural corrosion health assessment, it is thus crucial to understand how corrosion initiates from the microstructure to the component level and how structure corrosion behaviors change because of varied environmental stress factors. Many research efforts have been reported in the past addressing this critical issue [4, 8–10]. Traditionally, conventional ultrasonics and eddy current techniques have been used to precisely measure the thickness reduction in aircraft and other structures.

Besides, advanced corrosion health assessment systems require comprehensive quantitative information, which can be categorized into a variety of feature groups, such as corrosion morphology, texture, location, among others. It calls for the exploration of both new testing methods and data fusion methods from multiple testing techniques. Forsyth and Komorowski [11] discussed how data fusion could combine the information from multiple NDE techniques into an integrated form for structural modeling. Several other studies have looked into different sensing technologies for corrosion health monitoring, including using a micro-linear polarization resistance, μLPR sensor [12, 13], and fiber optic sensors [5]. However, the existing research effort in a combination of surface metrology and image processing is very limited. In parallel to the current corrosion sensing technology, there have been a number of corrosion modeling studies trying to numerically capture the processes of pitting corrosion initiation, pitting evolvement, pitting to cracking transition, and crack growth to fracture at the molecular level. However, currently there is no accepted quantitative model to take into consideration the effect of stress factors (e.g., salinity, temperature, pressure), although the effects of the above-mentioned stress factors have been widely discussed. Pit depth models of the form:

$$D(t) = \alpha(t - t_{ini})^\beta \text{ or } Dx/dt = g(x) = \beta\alpha^{1/\beta}x^{(1-1/\beta)} \text{ (pit growth rate)}$$

have been proposed in the literature.

Researchers proceed to describe then the transition from pit to crack (Kondo) and then the crack growth rate:

$$\frac{Dx}{dt} = C\sigma^p x^q$$

where σ is applied stress and C, p, and q are experimentally determined constants.

Our modeling methods are reasonable and compatible with our objectives, the data availability and the problem at hand. We would like to use Bayesian regression analysis to estimate the model parameters (stresses). The Bayesian technique can be combined with Markov chains-Monte Carlo sampling algorithms to arrive at the expression for each stress profile that will be included in the state model. Starting with $\underline{\theta} = \{p_1, p_2, \ldots\}$ and $Data = \{D_1, D_2, \ldots\}$ the probability of the parameter vector θ given the Data can be expressed as (Bayes' theorem):

$$\pi(\theta|Data) = \frac{L(Data|\underline{\theta})\pi_0(\theta)}{\int L(Data|\theta)\pi_0(\theta)d\theta}$$

where L is the likelihood function. We can assume initially that L is normally distributed.

Then the open source software WinBUGS can be used to give us the probabilistic profiles for the stresses. The data will be called upon to set the probability distributions initially.

For stress crack corrosion, the important stress factor is the <u>Stress Intensity Factor</u> (ΔK) Stress parameters will need to be normalized.

A metal surface (aluminum alloy, etc.) exposed to a corrosive environment may, under certain conditions experience attack at a number of isolated sites. The rate of dissolution in this situation is often much greater than that associated with uniform corrosion and structural failure may occur after a very short period. Several different modes of localized corrosion may be identified.

Corrosion modeling has been an active area of research over the recent past. A variety of models has been proposed depending on the material processes, the application domain, etc.

We address in this chapter analytical tools and methods to model accurately corrosion processes that are exploited eventually to design diagnostic and prognostic algorithms and assess the health state of critical nuclear waste storage facilities. The architecture is set as a decision support system providing advisories to the operator/maintainer as to the health status of such assets subjected to corrosion and in need of corrective action.

Of particular interest to our theme is localized corrosion and cracking, i.e. cracking initiating at points on the surface of a specimen (joints, fasteners, bolts, etc.). A metal surface (aluminum alloy, etc.) exposed to a corrosive environment may, under certain conditions, experience attack at a number of isolated sites. If the total area of these sites is much smaller than the surface area then the part is said to be experiencing localized corrosion. The rate of dissolution in this situation is often much greater than that associated with uniform corrosion and structural failure may occur after a very short period. Several different modes of localized corrosion may be identified. These are dependent on the type of specimen undergoing corrosion and its environment at the time of attack. Most destructive forms are pitting corrosion which is characterized by the presence of a number of small pits on the exposed metal surface, crevice attack and cracking. The rapidity with which localized corrosion can lead to the failure of a metal structure and the extreme unpredictability of the time and place of attack, has led to a great deal of study of this phenomenon. In this localized view, imaging studies are focusing on small areas of the global image where corrosion initiation is suspected and may spread more rapidly than other areas. We exploit novel image processing tools/methods, in combination with other means (mass loss calculations) to identify features of interest to be used in the modeling task, since imaging of corroding surfaces offers a viable, robust and accurate means to assess the extent of localized corrosion (see detailed treatment in the Data Analytics chapter).

We take advantage of first principle, semi-empirical and empirical approaches.

Most empirical/semi-empirical methods for corrosion fatigue modeling start with Paris' Law:

$$da/dN = C\Delta K^n$$

where da/dN is the rate of crack growth per cycle (m/cycle), C and n are empirical parameters, and ΔK is the stress intensity range. Micrographs of pitting and cracking corrosion are shown in Fig. 6.18.

Y. Kondo's corrosion modeling contributions are often quoted and employed by various investigators. In his original work, Kawai and Kasai [14] considered a three-stage corrosion process for industrial machines, i.e. pit growth, and crack formation from pit and fatigue crack propagation. He stipulated that pit volume for a hemispherical pit increases proportionally to time (t).

Although Kondo did not address directly aluminum alloy corrosion for aircraft structures, his work provided an empirical methodology to stipulate a modeling approach for crack formation from pit and to estimate a residual life for fatigue crack initiation based on inspection data.

Kawai and Kasai [14] investigated the fatigue crack initiation behavior of low-alloy steel in 90 °C de-ionized water. From this study, it was observed that the corrosion fatigue process is composed of three stages, namely, pit growth, crack formation from the pit, and corrosion fatigue crack propagation. The graph shown in Fig. 6.19 represents stages of the corrosion fatigue process highlighting crack growth rates. It was observed that the pit size increased with time (t) following the relation: pit size $\propto t^{1/3}$. The crack formation from the pit was determined from the stress intensity factor (linear elastic fracture mechanics parameter), which was calculated by assuming that the pit was a sharp crack.

The final model for the specimen under study by Kawai and Kasai [14] is of the form:

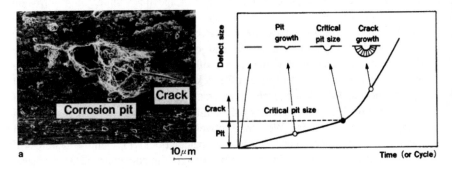

Fig. 6.18 Micrographs of pitting and cracking corrosion; evolution of the corrosion processes [14]

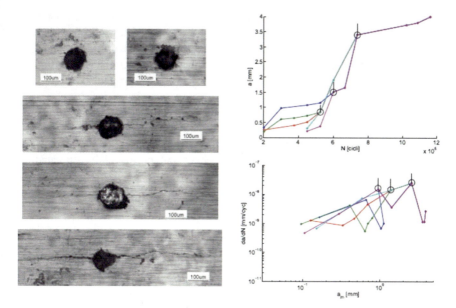

Fig. 6.19 Crack growth rate measurements

$$dC/dN = \left(\frac{1}{3}\right)C_p f^{-\frac{1}{3}} N^{-\frac{2}{3}} = \left(\frac{1}{3}\right)C_p^3 f^{-1} C^{-2}$$

Pit depth models of the form:

$$p(t) = \alpha(t - t_{ini})^\beta \text{ or } Dx/dt = g(x) = \beta \alpha^{1/\beta} x^{(1-1/\beta)} \text{ (pit growth rate)}$$

have been proposed in the literature.

Researchers proceed to describe then the transition from pit to crack (Kondo) and then the crack growth rate:

$$\frac{Dx}{dt} = C\sigma^p x^q$$

where σ is applied stress and C, p, and q are experimentally determined constants.

6.11 From Microscale to Mesoscale and Macroscale Models

Microscale Models At the atomic scale and based on the constituent materials' electrochemistry. Thermodynamic principles for corrosion attempt to explain why corrosion occurs. Aluminum alloy chemistry and the formation of Grain Boundaries (GBs) in such alloys address fundamental issues/concepts of how alloying elements (micro-constituents) add to strengthening and corrosion resistance. GBs contribute initially to structure defects and the formation of corrosion products at the microscale level. Chemical reactions at the GB contribute to pit growth.

Mesoscale Models At the mesoscale, measurements of mass loss and imaging results will be considered for modeling purposes when pitting becomes significant.

Macroscale Models Cracking is the basic corrosion mechanism at this stage where further action might be required when detected and its extent is estimated via prognostic methods.

Global Versus Local Corrosion Models It must be pointed out that initial modeling efforts, in conformance with current testing procedures and acquired data, will focus on global approaches, i.e. the whole panel area is viewed as the target for data collection and analysis. In contrast with a localized view where imaging studies, for example, are focusing on small areas of the global image where corrosion initiation is suspected and may spread more rapidly than other areas. This study will require novel image processing tools/methods to identify features of interest.

6.12 Corrosion Modeling Methods

Several recent books and reports address issues of corrosion modeling and describe thorough studies on all aspects of corrosion staging for aluminum alloy structures. Most notably, the book on Aluminum Alloy Corrosion of Aircraft Structures, edited by Derose and Suter [10] presents the results of work performed for the Simulation Based Corrosion Management (SICOM) project. The project was conducted under the auspices of the European Sixth Framework Program and aimed to develop a multi-scale corrosion modeling concept with a wide range of potential applications in research, development, and industry. The book defines the parameters and describes techniques needed for modeling and simulation of aluminum alloy corrosion in aircraft environments at the microscopic, mesoscopic, and macroscopic scales. The corrosion models address pitting and intergranular corrosion (microscale) of Al alloys, crevice corrosion in occluded areas, such as joints (mesoscale), galvanic corrosion of aircraft structural elements (macroscale), as well as, the effect of surface protection methods (anodization, corrosion inhibitor release, clad layer, etc.). The book describes the electrochemical basis for the models, their numerical implementation, and experimental validation and how the corrosion rate of the Al alloys at the various scales is influenced by its material properties and the surface protection methods. It will be of interest to scientists and engineers interested in corrosion modeling, aircraft corrosion, corrosion of other types of vehicle structures such as automobiles and ground vehicles, electrochemistry of corrosion, galvanic corrosion, crevice corrosion, and intergranular corrosion. Figure 6.20 depicts a general corrosion modeling approach.

Model validation is a necessary requirement in the corrosion assessment process. Software routines must be verified and the integrated model validated in order to be useful with high confidence. Figure 6.21 shows a path to model verification and validation.

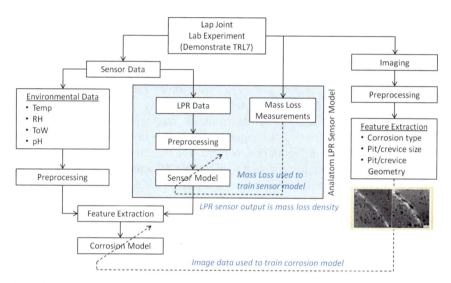

Fig. 6.20 Corrosion modeling approach

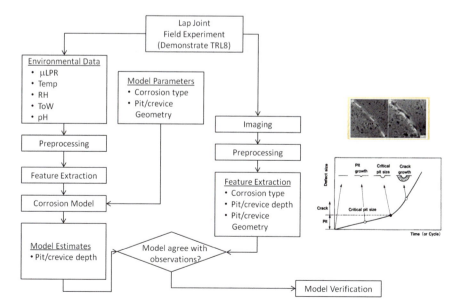

Fig. 6.21 Model validation

6.13 Data-Driven Models

Data driven corrosion models typically take advantage of constructs such as Artificial Neural Networks (ANNs), fuzzy and neuro-fuzzy systems and regression tools.

Investigators describe a computational approach using Wavelet transforms and artificial neural networks to analyze and quantify the extent of corrosion damage from NDI images. The Wavelet parameters obtained from the images were first used to classify between corroded and un-corroded regions using a clustering algorithm. The corroded regions were further analyzed to obtain the material loss due to corrosion using an artificial neural network model.

A cellular automaton (CA) modeling approach has been employed as the modeling medium to represent transitioning dynamics and to assist in the development of models for the corrosion stages.

Other approaches focus on the development and experimental validation of computational models for simulating galvanic corrosion in specific application case scenarios appearing in an aircraft environment. The numerical approach is based on solving the electro-neutrality equation with a **three dimensional Boundary/Finite Element Method**. Amongst the inputs of the problem are: geometrical description and physical properties of the electrolyte, as well as macroscopic polarization curves of the active electrodes. The main outcomes of the model are electric current density and potential distribution on the surface.

6.14 Model-Based Approaches

These approaches take advantage of first principle, semi-empirical and empirical modeling approaches to describe the initiation and evolution of corrosion processes. Several investigators describe the electrochemical basis for the models, their numerical implementation, and experimental validation and how the corrosion rate of the Al alloys at the various scales is influenced by its material properties and the surface protection methods. The corrosion models address **pitting** and **intergranular corrosion** (microscale) of Al alloys, **crevice corrosion** in occluded areas, such as joints (mesoscale), **galvanic corrosion** of aircraft structural elements (macroscale), as well as, the effect of surface protection methods (anodization, corrosion inhibitor release, clad layer, etc.).

6.15 Stochastic/Probabilistic Methods

The stochastic nature of pitting corrosion of metallic structures has been widely recognized. It is assumed that this kind of deterioration retains no memory of the past, so only the current state of the damage influences its future development. This characteristic allows pitting corrosion to be categorized as a **Markov process**. Localized corrosion, specifically pitting corrosion of metals and alloys, constitutes one of the main failure mechanisms of corroding structures such as aircraft components. Localized corrosion cannot be explained without assuming stochastic points of view due to the large scatter in the measurable parameters such as corrosion rate, maximum pit depth, and time to perforation [15]. Many variables of the metal-environment system such as alloy composition and microstructure, and composition of the surrounding media and temperature, are all involved in the pitting process [14]. Such complexity imposes the development of theoretical models and simulation tools for a better understanding of the outcome of the pitting corrosion process. These tools help predict more accurately the time evolution of pit depth in corroding structures as the key factor in structural reliability assessment. Another important characteristic of the pitting corrosion process that is worth noting is the time and **pit-depth** dependence of the corrosion rate [6, 7]. It has been established that, for a given pit, the growth rate decreases with time, while for pits with equal lifetimes, the corrosion rate is larger for deeper ones. Provan and Rodriguez [16] are amongst the first authors to use a nonhomogenous Markov process to model pit depth growth. In their model, the authors divided the space of possible pit depths into discrete, non-overlapping states and numerically solved the system of Kolmogorov's forward equations for the transition probabilities between damage states.

In recent years, modeling of pitting corrosion with **Markov chains** has shown new advances. For example, Bolzoni et al. [17] have modeled the first stages of localized corrosion using a continuous-time, three-state Markov process. The Markov states of the metal surface are passivity, meta-stability, and stable pit growth. On the other hand, Timashev et al. [18] formulated a model based on the use of a continuous-time, discrete-state pure birth homogenous Markov process for stochastically describing the growth of corrosion-caused metal loss. In their model, the intensities of the process were calculated by iteratively solving the proposed system of Kolmogorov's forward equations.

The stochastic nature of pitting corrosion of metallic structures has been widely recognized. It is assumed that this kind of deterioration retains no memory of the past, so only the current state of the damage influences its future development. This characteristic allows pitting corrosion to be categorized as a Markov process. In this paper, two different models of pitting corrosion, developed using Markov chains, are presented. Firstly, a continuous-time, nonhomogeneous linear growth (pure birth) Markov process is used to model external pitting corrosion in underground pipelines. A closed-form solution of the system of Kolmogorov's forward equations is used to describe the transition probability function in a discrete pit depth space. The transition probability function is identified by correlating the stochastic pit depth mean with the empirical deterministic mean. In the second model, the distribution of maximum pit depths in a pitting experiment is successfully modeled after the combination of two stochastic processes: pit initiation and pit growth. Pit generation is modeled as a nonhomogeneous Poisson process, in which induction time is simulated as the realization of a Weibull process. Pit growth is simulated using a nonhomogeneous Markov process. An analytical solution of Kolmogorov's system of equations is also found for the transition probabilities from the first Markov state. Extreme value statistics is employed to find the distribution of maximum pit depths.

López De La Cruz et al. [19] proposed a stochastic analysis method of spatial point patterns as effect of localized pitting corrosion. A test of randomness is performed by means of the inter-event distance method. The robustness of the method is tested with two artificially generated samples. The method is applied to published empirical data. The results complied when spatial regularity of pits was found and showed discrepancy when pit interaction was observed.

Probabilistic modeling includes all significant uncertainties that affect aircraft component reliability, such as flight conditions, operational loading and environmental severity, manufacturing deviations, material properties and maintenance inspection activities. Advanced response surface modeling tools based on stochastic field approximation models are employed for computing the local bivariate stochastic stresses (mean stress and stress range are the two correlated stress components). The ProCORFA software developed by GP Technologies in collaboration with STI Technologies for USAF was employed in several such studies.

A three-level hierarchical, multi-scale, **stochastic FE analysis** approach was proposed and employed for computing stochastic local stresses. The employed three-level hierarchical stochastic FE analysis is capable of computing accurately the stochastic stress variations near rivets that are caused by loading, material, geometric configuration uncertainties, including deviations from the baseline geometry of the lap joint due to manufacturing process.

At the top level, a global airframe FE model with a relative coarse mesh was used. At an intermediate level, a local FE model of the lapjoint was used. The computed displacement response of the global FE model was considered to be the input boundary conditions for the local FE analysis of the lap joint. This local FE model included the joint rivets and splices, plus the contact surface conditions between the joint components. At the bottom level, a local axisymmetric FE model of a single rivet was employed. This detailed single rivet FE model has a very refined mesh that was required for incorporating accurately material plasticity effects on the local contact stresses around the rivet.

Other investigators suggested a finite element method (FEM)-based corrosion model, specifically tailored for localized pitting corrosion of aluminum alloys. The model distinguishes itself from existing ones by its strong predictive power and high generality. By resorting to this methodology, not only corrosion rate but also pit stability can be quantitatively evaluated for a wide range of systems involving heterogeneous alloy microstructure, complex pit morphology, and versatile solution chemistry.

The stochastic theory of pitting corrosion has been successfully used to analyze the statistical nature of pitting. A Monte Carlo model has been proposed to predict the extent of damage accumulation in aluminum alloys. This model uses experimental parameters obtained by electrochemical noise measurements on electrode arrays. The algorithm is based on the random occurrence of the metastable pit birth/death or the stable pit growth. Simulated pit depth distributions are compared to experimental data obtained by Optical Profilometry (OP), leading to an improvement of the model and challenging the existence of a metastable/stable transition in free corrosion conditions. The evolution of metastable pit birth rate with time shows an initial linear increase and an exponential decay. Monte Carlo modeling can successfully reproduce experimental pit depth distributions in aluminum alloys.

A probability approach for life prediction is developed and illustrated in the work of Harlow and Wei [20] through a simplified model for the pitting corrosion and corrosion fatigue crack growth in aluminum alloys in aqueous environments. A method for estimation of the cumulative distribution function (CDF) for the lifetime is demonstrated by using an assumed CDF for each key random variable (RV). The basic aim of this approach is to make predictions for the lifetime, reliability, and durability beyond the range of typical data by integrating the CDFs of the individual RVs into a mechanistically based model. The contribution of each key RV is considered, and its significance is assessed. Thus, the usefulness of probability-based modeling is demonstrated. It is noted that physically realistic parameters were assumed for the illustrations. As such, the results from analysis of the model qualitatively agree quite well with experimental observations. However, these results should not be construed to represent behavior in actual systems. Because of these assumptions, confidence levels for the predictions are not addressed.

6.16 Corrosion Modeling Approaches

Corrosion in metal structures typically proceeds from pitting to cracking, as shown schematically in Fig. 6.22. Characteristic corrosion stages are shown in Fig. 6.23.

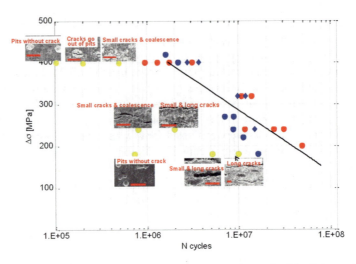

Fig. 6.22 From pitting to cracking of corroding specimens (*Source* Dr. Vinod Agarwala)

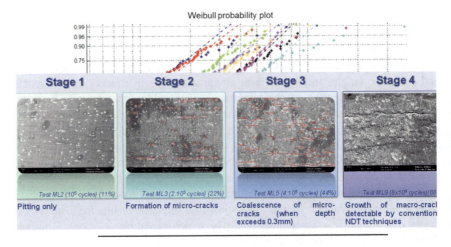

Fig. 6.23 Corrosion stages from pitting to cracking; overlapping stages with a Weibull probability plot

The tables below are a small sample of studies reported over the past years on corrosion fatigue mechanisms, pitting and cracking issues and modeling approaches (Table 6.2 and 6.3).

6 Corrosion Modeling

Table 6.2 Modeling pitting corrosion fatigue: pit growth and pit/crack transition issues [1]

	Proposed by	Summary	Description	Advantages/limitations
1	Hoeppner [21–current]	• Proposed a model to determine critical pit depth to nucleate a Mode I crack under pitting corrosion fatigue conditions • Combined with the pit growth rate theory as well as the fatigue crack growth curve fit in a corrosive environment, the cycles needed to develop a critical pit size that will form a Mode I fatigue crack can be estimated	• Using a four parameter Weibull fit, fatigue crack growth threshold (ΔK_{th}) was found from corrosion fatigue experiments for the particular environment, material, frequency, and load spectrum • The stress intensity relation for surface discontinuity (half penny shaped crack) was used to simulate hemispherical pit, i.e., $$K = 1.1\sigma\sqrt{\pi\left(\frac{a}{Q}\right)}$$ where, σ is the applied stress, a is the pit length, and Q is the function of a/2c, S_{ty} • Using the threshold determined empirically, critical pit depth was found from the stress intensity relation mentioned above • Then, the time to attain the pit depth for the corresponding threshold value was found using: $$t = \left(\frac{d}{c}\right)^3$$ where, t is the time; d is the pit depth, and c is a material/environment parameter	• This model provides a reasonable estimate for hemispherical geometry of the pits • This model is useful to estimate the total corrosion fatigue life with knowledge of the kinetics of pitting corrosion and fatigue crack growth • This model did not attempt to propose mechanisms of crack nucleation from corrosion pits • This model is valid only for the conditions in which LEFM concepts are applicable • Material dependent

(continued)

Table 6.2 (continued)

	Proposed by	Summary	Description	Advantages/limitations
2	Lindley et al. [22]	• Similar to Hoeppner's model, a method for determining the threshold at which fatigue cracks would grow from the pits was proposed	• Pits were considered as semi-elliptical shaped sharp cracks • Used Irwin's stress intensity solution for an elliptical crack m an infinite plate and came up with the relationship to estimate threshold stress intensity values related to fatigue crack nucleation at corrosion pits, i.e., $$\Delta K_{th} = \frac{\Delta \sigma \sqrt{(\pi a)} \left[1.13 - 0.07(a/c)^{1/2}\right]}{\left[1 + 1.47(a/c)^{1.64}\right]^{1/2}}$$ where, $\Delta \sigma$ is the stress range, a is the minor axis, and c is the major axis of a semi-elliptical crack • From the observed pit geometry, i.e., for a/c ratio, threshold stress intensity can be calculated • For the corresponding a/c ratio, critical pit depth can be estimated	• The proposed stress intensity relation can be used in tension—tension loading situations where stress intensity for pits and cracks are similar • Critical pit depths for cracked specimens can be estimated using the existing threshold stress intensity values • This model is valid only for the conditions in which LEFM concepts are applicable • Material dependent
3	Kawai and Kasai [14]	• Proposed a model based on estimation of allowable stresses under corrosion fatigue conditions with emphasis on pitting • As corrosion is not usually considered in developing S-N fatigue curves, a model for allowable stress intensity threshold involving corrosion fatigue conditions was proposed	• Considered corrosion pit as an elliptical crack • Based on experimental data generated on stainless steel, new allowable stresses based on allowable stress intensity threshold was proposed, i.e., $$\Delta \sigma_{all} = \frac{\Delta K_{all}}{F \sqrt{\pi h_{max}}}$$ where, ΔK_{all} can be determined from a da/dN versus ΔK plot for a material, hmax is the maximum pit depth, and F is a geometric factor	• Using this model, allowable stress in relation to corrosion fatigue threshold as a function of time can be estimated • Material dependent • This model is valid only for the conditions in which LEFM concepts are applicable

6 Corrosion Modeling

Table 6.2 (continued)

	Proposed by	Summary	Description	Advantages/limitations
4	Kondo [23]	• Corrosion fatigue life of a material could be determined by estimating the critical pit condition using stress intensity factor relation as well as the pit growth rate relation	• Pit diameter was measured intermittently during corrosion fatigue tests • From test results, corrosion pit growth law was expressed as $2c \propto C_p t^{1/3}$ where, 2c is the pit diameter, t is the time, and C_p is an environment material parameter Then, critical pit condition (ΔK_p) in terms of stress intensity factor was proposed by assuming pit as a crack $\Delta K_p = 2.24\sigma_a \sqrt{\pi c \, \alpha/Q}$ where, σ_a is the stress amplitude, a is the aspect ratio, and Q is the shape factor • Critical pit condition was determined by the relationship between the pit growth rate theory and fatigue crack growth rates $c = cp(N/f)^{1/3}$ where, N is the number of stress cycles, f is the frequency, and 2c is the pit diameter • The pit growth rate dc/dN was developed using ΔK relation as given below $dc/dN = \left(\frac{1}{3}\right) C_p^3 f^{-1} \alpha^2 \pi^2 Q^{-2} (2.24\sigma_a)^4 \Delta K^{-1}$ dc/dN was determined using experimental parameter C_p • Finally, the critical pit size $2C_{cr}$ was calculated from the stress intensity factor relation i.e., $2C_{cr} = (2Q/\pi\alpha)(\Delta K_p/2.24\sigma_a)^2$	• The aspect ratio was assumed as constant • Material and environment dependent

Table 6.3 Corrosion modeling approaches

Corrosion modeling approaches in tile literature	
Paris' Law	$\frac{da}{dN} = C\Delta K^M$
Kotzalas and Harris	$F\Delta t - \frac{m\pi d_m n(1-\gamma)^2}{60D} x$
Choi and Liu	$\frac{da}{dN} = C \cdot \frac{H_b}{H_1} (\Delta K)^n$
Kondo	$2c \alpha C_p t^{\frac{1}{3}}$
Harlow and Wei	The number of cathodic particles in a cluster is a random variable with discrete Pareto distribution
	$p_k = P(N_c = k) = 0.725 \quad k^{-2.41} \quad k \geq 1$
	The pit grows at a constant volumetric rate according to Faraday's Law:
	$\frac{dV}{dt} = \frac{MI_p}{nF\rho} = \frac{MI_{p0}(k)}{nF\rho} \exp(-\frac{\Delta H}{RT})$

6.17 Global Versus Local Corrosion Models

A metal surface (aluminum alloy, etc.) that is exposed to environmental stresses may experience a corrosive attack at a number of isolated sites. A part is experiencing localized corrosion if the total area of these sites is much smaller than the total surface area [22]. Several different modes of localized corrosion may occur such as pitting and crevice attack. Pitting corrosion is characterized by the presence of a number of small pits on the exposed metal surface. The geometries of the pits depend on many factors such as the metal composition and the surface orientation [22]. Crevice attack occurs in situations where two or more surfaces in close proximity lead to the creation of a locally occluded region in which enhanced dissolution may occur. Figure 6.24 is an illustration of main corrosion forms.

Detection, localization and sizing of corrosion in complex structures over large, partially accessible areas are of growing interest in the aerospace industries. Traditionally, conventional ultrasonic thickness gauging and eddy current techniques have been used to precisely measure the thickness in structures. However, the scanning may become impossible when the area of inspection is inaccessible. Upon this need, there has been a number of undergoing research using guided wave tomography techniques to screen large areas of complex structure for corrosion detection, localization [7] and defect depth mapping [8]. However, due to the nature of ultrasonic guided wave, this technique is vulnerable to environmental changes, especially to temperature variation and surface wetness, and the precision of corrosion defect thickness reconstruction is restricted by sensor network layout, structure complexity and other factors, which limits the scope of the field application.

It must be pointed out that initial modeling efforts, in conformance with current testing procedures and acquired data, focus on global approaches, i.e. the whole panel area is viewed as the target for data collection and analysis. In contrast with a localized view where imaging studies, for example, are focusing on small areas of

6 Corrosion Modeling

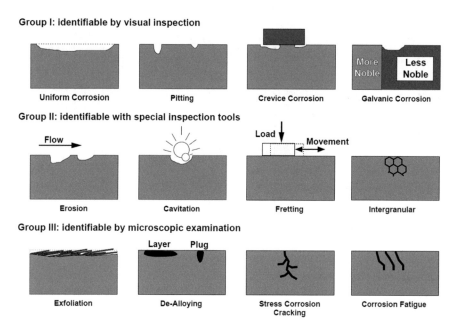

Fig. 6.24 Main forms of corrosion regrouped by their ease of recognition [6]

the global image where corrosion initiation is suspected and may spread more rapidly than other areas. This study requires novel image processing tools/methods to identify features of interest.

It is evident that imaging of corroding surfaces offers a viable, robust and accurate means to assess the extent of localized corrosion. Modeling, in combination with sensor measurements, promises to assist in the timely and accurate corrosion detection and prediction.

Preliminary results from a global perspective derived from current analysis efforts are listed in the sequel.

6.18 A Novel Modeling Approach

It is obvious from the discussion above that the fundamental corrosion mechanisms of initial corrosion, pitting and cracking involve electrochemical and physical processes that require a separate modeling perspective for each, as reflected also on the testing, data collection and analysis procedures. As an example of the suggested modeling framework, we present briefly a modeling study conducted by Georgia Tech under sponsorship by the US Army for corrosion detection and prediction of rolling element bearings. Figure 6.25 depicts the scheme and modeling results computed for corrosion detection and prediction of rolling element bearings.

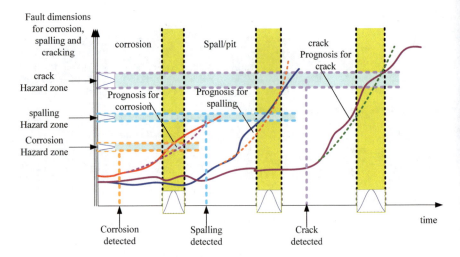

Fig. 6.25 Multi-fault corrosion modes

The corrosion mechanism here, characteristic of a rolling element bearing under dynamic loading are corrosion, spalling and cracking. It is suggested only as an example of the intended modeling framework.

It is evident that the three corrosion mechanisms considered in this study progress at different rates and, therefore, require separate modeling approaches. The fault progression is often nonlinear and, consequently, the model should be nonlinear. From a nonlinear Bayesian state estimation standpoint, diagnosis and prognosis may be accomplished by the use of a Particle Filter-based module. An essential element of this module is a nonlinear state model describing the propagation of the fault.

From the above descriptions, it is clear that there are usually more than one fault modes present in a corroding specimen/panel. Although a corrosion model can describe the propagation of corrosion, a pitting model can describe the propagation of pitting, and a cracking model can describe the propagation of cracking, an extended model to describe a "combined" or "fused" health condition is desired. Since these different fault modes work together we can write

$$\frac{dg}{dR} = f[b_1 w_1 (corrosion) + b_2 w_2 (pitting) + b_3 w_3 (cracking)]$$

where w_1, w_2, w_3 ($w_1 + w_2 + w_3 = 1$) are weighting factors for different fault modes, b_1, b_2, b_3 are time varying parameters indicating that corrosion, pitting, and cracking are detected and their respective prognostic module is activated, respectively. The crack progression model can be written in the form:

6 Corrosion Modeling

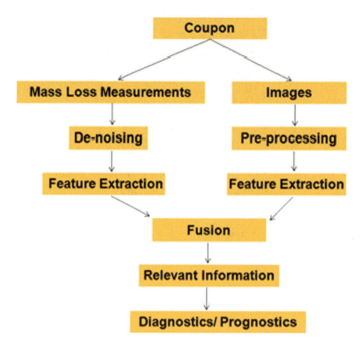

Fig. 6.26 Basic modules of the corrosion modeling architecture: from measurements to corrosion detection and prediction

$$D(t+1) = D(t) + C_D(D(t))^{n\left(w_h h_i + w_t t_i + w_l l_i + w_g g_i\right)}$$

where the term C captures the parameters influencing the crack growth progression and may be time-varying.

We introduce a combined model-based and data-driven methodology that takes advantage of experimental data, corrosion progression models and an estimation method called particle filtering. It is accompanied by appropriate performance metrics while meeting customer specified requirements for detection confidence and false alarm rates, and prediction accuracy/precision. Figure 6.26 shows the basic modules of the corrosion modeling framework.

6.19 A General Framework to Corrosion Modeling

$$L(k+1) = L(k) + C\alpha(k)\{\Delta K^m\} + v_1(k)$$
$$\uparrow$$
$$\text{Stress intensity factor range}$$

or

$$L(k+1) = L(k) + C\alpha(k)\{\Delta K^m + p(k)\} + v_1(k)$$
$$\alpha(k+1) = \alpha(k) + v_2(k)$$
$$\uparrow$$
Un-modeled

$$\Delta K(k) = f(load, L(k))$$

$$p(k) = p_1(k), p_2(k), p_3(k), p_4(k)\ldots = Salinity, RH, pH, Time\ of\ wetness\ etc.$$
$$Feature(k) = L(k) + n(k)$$
or
$$Feature(k) = h(L(k)) + n(k)$$

where v_1, v_2, n are noise profiles, C is a parameter of the material.

6.20 Other Failure Prediction Models

Alternative model structures are available and may be appropriate under particular fault model conditions. The corrosion community has used Kondo's model, shown in Fig. 6.27, extensively.

Kondo's model:

$$dC/dN = \left(\frac{1}{3}\right) C_p f^{-\frac{1}{3}} N^{-\frac{2}{3}} = \left(\frac{1}{3}\right) C_p^3 f^{-1} C^{-2}$$

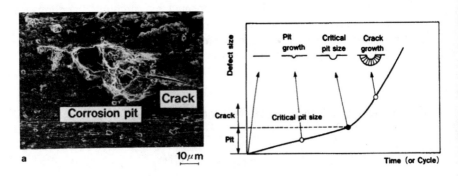

Fig. 6.27 Kondo's model

6 Corrosion Modeling

Pit depth:

$$d(t) = \left(\frac{3m_{loss}(t)}{2\pi\rho}\right)^{\frac{1}{3}}$$

A modeling framework has been proposed for pitting of stainless steel canisters by the following formulation:

$$f(x; k, \lambda, \theta) = \frac{k}{\lambda}\left(\frac{x-\theta}{\lambda}\right)^{k-1} \exp\left(-\frac{x-\theta}{\lambda}\right)^{k}$$

where x is the pit depth, lambda is the average pit diameter, and k characterizes the spread of average pit diameters.

Pit size distribution changes as functions of time:

$$x = \alpha t^{\beta}$$

Alpha and beta are determined experimentally.

Pit growth rate:

$$\frac{dx}{dt} = g(x) = \beta \alpha^{1/\beta} x^{(1-1/\beta)}$$

6.21 Stochastic Dynamical Model of Corrosion States from Pitting to Cracking Under Loading and Environmental Stress

For the past decades, pitting corrosion and cracking can lead to great mass loss and result in decreased product performance. Thus, it is essential to come up with a methodology to assess and predict corrosion status. Here modeling is the foundation of giving an assessment of status and prediction. However, currently there is not a widely accepted model, which can take the effect of the stress factors (e.g. salinity, temperature, pressure, etc.) into consideration. The effects of these stress factors have been discussed but literature is scarce as to what exactly those functions might be. We discuss in this section the effect of the environmental stress factors on the corrosion rate.

6.22 Pitting Corrosion

A conceptual model of corrosion growth could track the depths [mm] from uniform corrosion (surface loss) in addition to localized corrosion (e.g., pitting and crevices)

$$d(x,t) = d_{uniform}(t) + d_{localized}(x,t)$$

as presented by Straub [18], however, no practicable framework currently exists for detailed spatial information x (in 1-, 2-, or 3D). Thus, the focus turns to the maximum depth $d_{max}(t)$ over a region (e.g., deepest pit in a whole pipe or panel), whose statistics can be treated with Gumbel or Weibull extreme value distributions.

Data from atmospheric uniform corrosion support use of a power growth over time:

$$d(t) = at^b$$

where a, b parameters are functions of temperature, relative humidity (RH), time of wetness (ToW), salinity, SO_2, etc., and values can vary significantly only meters apart due to different microenvironments. The same power form applies to localized corrosion, however because anodes (pits) are smaller than the surrounding cathodic areas, pitting damage can grow considerably faster than uniform damage.

From Kondo's experiments, pit volume (or material loss [mg]) tends to increase linearly, so pit depth grows with $t^{1/3}$. That 1/3 is our a priori b. Furthermore, depth-to-radius ratio tends to stay constant around 0.7, so radius $R(t)$ or area $A(t)$ features measured at time t can map to a hidden depth state assuming a hemispherical or semi-ellipsoidal pit. For our model, we will be interested in the inverse relation

$$A_{measured}(t) = \pi(1.43 \cdot d(t))^2 + noise$$

where we took the observed radius of an equivalent circle (top view of a pit) to be 0.7 $R(t)$.

With uncertainty around the mean also expected to grow at this same power of t, the time-varying Gumbel pdf for $d_{max}(t)$ is

$$p(d,t) = \frac{1}{\beta(t)} e^{-\left[\frac{d-\mu(t)}{\beta(t)} + e^{-\left[\frac{d-\mu(t)}{\beta(t)}\right]}\right]}$$

where $\mu = at^b$ (mean parameter) and $\beta = \beta_0 t^b$ (scale parameter). We note that there would be a problem drawing from this distribution over time at the same location: maximum pit depth cannot go backwards (smaller). Therefore, this distribution is better suited to representing samples of d_{max} over a partitioned space as suggested by [18], with respect to pits further apart than a correlation length, and still behaving as $t^{1/3}$ over time.

6.23 Paris' Law Revisited

The classic Paris fatigue crack growth rate for metals is a power law

$$\frac{dL}{dN} = C\Delta K^m$$

where L = crack size (length or depth), N = number of cycles (like a usage-based time variable), $\Delta K = K_{max} - K_{min} \left[Pa\sqrt{m} = Nm^{-\frac{3}{2}} \right]$ is stress intensity factor range, and C, m are empirical parameters associated to a material. This relation looks like a straight line in log-log scales, with slope m and intercept C. Stress intensity factor K measures how "concentrated" stresses are around the crack tip. K is proportional to stress load amplitude, $\sqrt{L}$ (or its reciprocal), and a dimensionless correction for geometry $Y(L)$ (e.g., boundaries). For example, a typical situation is a center crack with remote stress applied uniformly:

$$\Delta K = \Delta S \sqrt{\pi L} \cdot Y(L)$$

where $\Delta S = S_{max} - S_{min}$ over one cycle. ΔK may also have a correction for crack closure effect (retarding growth). What Paris says is that for a midrange of ΔK values, valid only during the crack propagation phase, between crack initiation and fracture, a crack will grow at the rate predicted by the straight line. (Growth is still very slow, 1e-4 to 1e-2 mm/cy, and striations may be seen in this phase.) The higher the applied maximum stress, the faster the path to failure, and bigger crack size can beget even faster rate (or can slow down if stress is wedged from the inside so that crack tip gets farther away from the stress as it grows.) Regardless of how fast or slow, the positive value of rate says that the specimen will eventually fail, especially since its physical dimensions are finite.

For metals, m tends to be between 2 and 4 (e.g., 3.4 steel, 2.85 aluminum alloy), though range is wider. Literature and published data support interdependence

$$lnC = -15.84 - 3.34m$$

which is expected since the smaller (flatter) the positive slope, the higher the intercept will be in log-log scales.

The parameters C, m depends on material, environmental conditions (e.g., temperature), and to some extend stress ratio ($\frac{S_{min}}{S_{max}} = \frac{K_{min}}{K_{max}}$) a bad proxy for mean stress over one cycle). Figure 6.28 shows the dependence of the C parameter on m. Growth rate curve can also change with frequency of cycles and even the ordering of stresses if loading profile varies. In summary, Paris' law captures a big kernel of truth about crack growth; while myriad other factors such as mean stress and the environment act as modulators or corrections to this basic law. The latter concept is

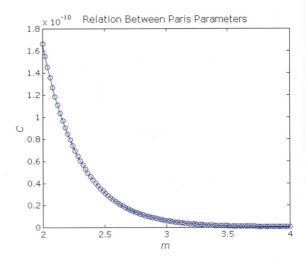

Fig. 6.28 The parameter C as a function of m

particularly important in corrosion modeling because, as we will see, catastrophic failure is predicted to occur much earlier if these modulating factors are taken into account.

6.24 Transition from Pitting to Cracking

"Nanocracks" (below fracture mechanics law) cannot be practicably inspected, but do happen prior to visible macrocracks. Phillips and Newman found that a dK analysis extending to the left of $\Delta K th$ would typically look convex (concave-up), but noted that attempting to model this subthreshold phenomenon with crack sizes starting below 0.3 mm does not make a difference. Thus, in our model we simply assume that an initial value of visible crack size suddenly appears at some elapsed cycle N_{init}. In one interpretation of our corrosion model, the sudden appearance of a crack happens during pitting growth and under cyclic loading. A threshold is eventually surpassed where stresses at tip(s) of the largest corrosion defect are concentrated enough for crack(s) to begin propagation along the plane (or thru wall). From that time on, the relevant maximum defect size becomes crack length (or depth) as opposed to pit depth, and is dominated by our complete sigmoidal dynamic law until eventual fracture. This motivates an integrated measure of multimodal fault as a weighted sum or norm of pit depth and crack length. In the present work we use infinity norm of equally-weighted faults as the *Combined Degradation Index*:

$$I_d(t) = \left\| [d_{pit}; L_{crack}] \right\|_\infty = \max\left(d_{pit}, L_{crack}\right)$$

6.25 Environmental Stressors

As mentioned before, several degradation parameters are functions of the environment, however, literature is scarce as to what exactly those functions might be. Empirical curves may be available for a few stresses or a couple of temperatures, but we seek relations partly based on the underlying materials science to have a better chance at generalizability.

The deWaards-Milliams model of carbonic acid (water + CO_2) corrosion in steel pipes is too crude for spatiotemporal prediction, however that is the "state of the art" and does capture a static relation between corrosion rate and the environmental stressors temperature, pressure, and fraction of acid in gas form

$$r = 10^{5.8 - \frac{1710}{T} + 0.67 \log f}, f = nP \cdot 10^{P(0.0031 - 1.4/T)}$$

where r is for the maximum thru-wall depth of damage (though unknown where in the pipe) [mm/yr], T = temperature [K], P = pressure [bar], n = fraction in gas. After algebra and changing Kelvin to degrees Celsius, the whole expression reduces to

$$r = (nP)^{0.67} \times 10^{5.8 - \frac{1710}{T+273.15} - \frac{0.93P}{T+273.15} + 0.0021P}$$

Expected working ranges are $T \sim 20-80°$ C (e.g., 30), $P \sim$ 50–200 bars (e.g., 100), $n = 0.01$. It is conjectured that such an expression, with tunable parameters, has better extrapolation prospects than purely black-box models when applied to other metals and situations.

The plot in Fig. 6.29 shows the dependence of corrosion rate on pressure and temperature.

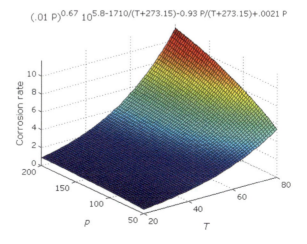

Fig. 6.29 Dependence of corrosion on pressure and temperature

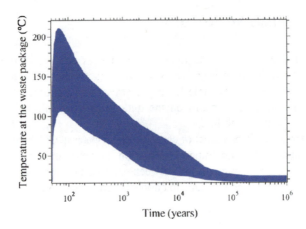

Fig. 6.30 Temperature at the waste package as a function of time

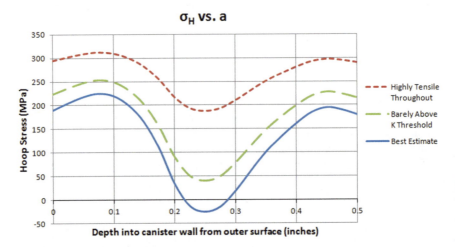

Fig. 6.31 Hoop stress as a function of depth into canister wall from outer surface

Figure 6.30 depicts the temperature profiles of the waste package as a function of time.

Figure 6.31 shows the hoop stress as a function of depth into the canister wall.

6.26 Symbolic Regression Modeling Framework

We exploit a novel modeling framework to represent stress and/or corrosion evolution in corroding structures. This modeling approach allows for inclusion of impact or stress factors while accounting for uncertainty.

6 Corrosion Modeling

Symbolic Regression is a type of regression analysis that searches the space of mathematical expressions to find the model that best fits a given dataset, both in terms of accuracy and simplicity. No particular model is provided as a starting point to the algorithm. Instead, initial expressions are formed by randomly combining mathematical building blocks such as mathematical operators, analytic functions, constants, and state variables. (Usually, the person operating it will specify a subset of these primitives, but that is not a requirement of the technique.) New equations are then formed by recombining previous equations, using genetic programming.

By not requiring a specific model to be specified, symbolic regression isn't affected by human bias, or unknown gaps in domain knowledge. It attempts to uncover the intrinsic relationships of the dataset, by letting the patterns in the data itself reveal the appropriate models, rather than imposing a model structure that is deemed mathematically tractable from a human perspective. The fitness function that drives the evolution of the models takes into account not only error metrics (to ensure the models accurately predict the data), but also special complexity measures, thus ensuring that the resulting models reveal the data's underlying structure in a way that's understandable from a human perspective. This facilitates reasoning and favors the odds of getting insights about the data-generating system (abstracted from Wikipedia).

- Quick Recall: regression is aiming to minimize the distance between the estimated value and measured value:

$$arg \min_{f}(y - f(x))^2$$

Trying to find the best g(t).

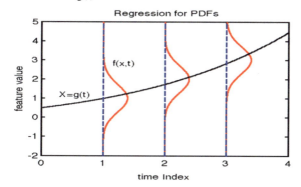

Objective Function:

For each t_i, the distance (error) can be represented as:

$$E_i = \int_{x=-\infty}^{+\infty} (g(t_i) - x)^2 f(x, t_i) dx, \quad i = 1, 2, \ldots N$$

Therefore, the objective function is the sum of E_i:

$$\arg\min_g E = \sum_i E_i = \sum_{i=1}^{N} \int_{x=-\infty}^{+\infty} (g(t_i) - x)^2 f(x, t_i) dx$$

6.27 Discrete Form

- We can generate samples using Monte Carlo Methods and change the distributions to samples.
- First, we assume for each time instant, t_i, the numbers of sample are the same.
- We want to generate the corresponding discrete form, which can fit our coding easier.

$$\arg\min_g E = \sum_i E_i = \sum_{i=1}^{N} \sum_{k=1}^{M} (X_{t_i,k} - g(t_i))^2$$

- Here M is the number of sample for each distribution.
- When the number of samples for each distribution is not the same, we can use the histogram to approximate the distributions.
- The objective function is changed to the following form:

$$\arg\min_g E = \sum_i E_i = \sum_{i=1}^{N} \frac{\sum_{k=1}^{M_i} (X_{t_i,k} - g(t_i))^2}{M_i} = \sum_{i=1}^{N} \sum_{k=1}^{M_i} \left(\frac{X_{t_i,k} - g(t_i)}{\sqrt{M_i}}\right)^2$$

M_i is the number of samples for the distribution at time t_i.

6.28 Useful Tools

- Eureka® is a useful tool to finish the Symbolic Regression:

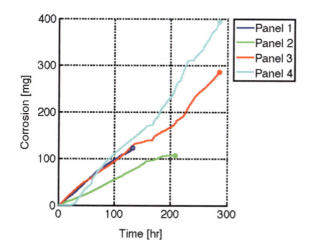

Sample generating code:

```
1   clc
2   clear all
3   load('C_data.mat')
4
5   N=15; % sample for each fitted
6   Eureqadata=[];
7
8   for i=1:360:length(time)
9       if time(i)<133
10          pd=fitdist(Corrosion(i,:)','normal');
11      elseif time(i)<209
12          pd=fitdist(Corrosion(i,2:4)','normal');
13      else
14          pd=fitdist(Corrosion(i,3:4)','normal');
15      end
16      Eureqadata=[Eureqadata;time(i)*ones(N,1),normrnd(pd.mu,pd.sigma,[1 N])'];
17
18  end
19  figure
20  hold on
21  for i=1:length(time)/360*N
22      scatter(Eureqadata(i,1),Eureqadata(i,2));
23  end
24
25  save Eureqadata
26
```

Generated sample plot:

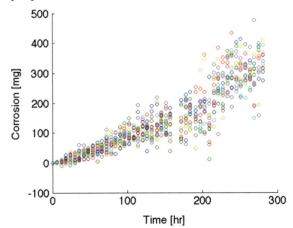

Fitting result:

- Find best solution.

Best Solutions of Different Sizes

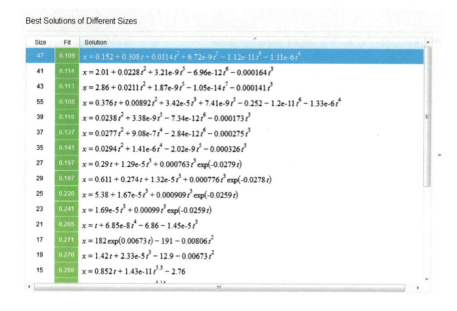

- We can choose Fit or Solution to minimize the error, or introduce a tradeoff between size and error.
- We carry out the performance analysis with the following function:

$$f1(t) = 0.152 + 0.308 * t + 0.0114 * t^2 + 6.72e - 9 * t^5 - 1.12e - 11 * t^6 - 1.11e - 6 * t^4$$
$$f2(t) = 2.01 + 0.0228 * t^2 + 3.21e - 9 * t^5 - 6.96e - 12 * t^6 - 0.000164 * t^3$$
$$f3(t) = 2.86 + 0.0211 * t^2 + 1.87e - 9 * t^5 - 1.05e - 14 * t^7 - 0.000141 * t^3$$
$$f4(t) = 182 * e^{(0.00673*t)} - 191 - 0.00806 * t^2$$

- We test these curves with our testing data:

Fitted curve	Error
F1(t)	1.0881
F2(t)	1.0710
F3(t)	1.1029
F4(t)	1.1482

6.29 An Example

This example exploits a set of pressure data:

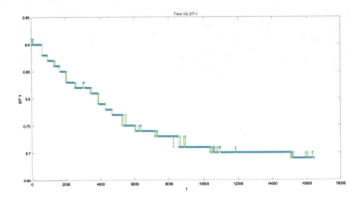

- Plot of Pressure Data.

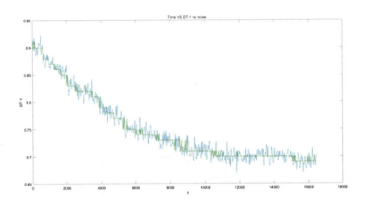

- Noise Added to the signal

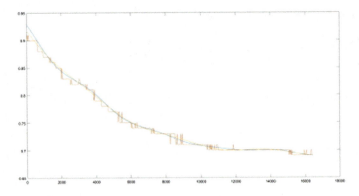

- Modeling (blue) vs Original (orange) Plot.

6.30 An Extended Corrosion Model

There usually more than one fault modes present in a corroding panel. Although a corrosion model can describe the propagation of corrosion, a pitting model can describe the propagation of pitting, and a cracking model can describe the propagation of cracking, an extended model to describe a "combined" or "fused" health condition is desired. Since these different fault modes work together in a coupled fashion, this "combined" or "fused" health condition could provide a more precise and accurate evaluation of the health of the panel.

To this end, a generalized degradation (fault) variable is defined as g. Then, the propagation of this generalized degradation variable can be described as

$$\frac{dg}{dR} = f[b_1 w_1 (corrosion) + b_2 w_2 (spalling) + b_3 w_3 (cracking)] \qquad (6.1)$$

where w_1, w_2, w_3 ($w_1 + w_2 + w_3 = 1$) are weighting factors for different fault modes, b_1, b_2, b_3 are time varying parameters indicating that corrosion, pitting, and cracking are detected and their respective prognostic module is activated, respectively.

To use this model in diagnosis and prognosis, we must consider the fault features and ground truth data. In Eq. (6.1), the different fault modes are weighted and summed. However, the features cannot be added directly. To solve this problem, the feature vectors for different fault modes can be fused by some intelligent methods, such as genetic algorithm, genetic programming, Dempster-Shafer theory, neural networks, fuzzy logic, Kalman filtering, etc.

6.30.1 Sensor Modeling Parameters

The corrosion rate (CR) is computed using the following formula,

$$CR = B_{corr}/R_p$$

where R_p is the measured polarization resistance and B_{corr} is a constant of proportionality made up of several physical parameters, physical constants and electrochemical properties of the μLPR sensor. A description of how B_{corr} is computed is provided in Fig. 6.32.

The Tafel coefficients, β_a and β_a, were acquired using a potentiostat (Gamry Reference 600). The μLPR was placed flush against a graphite counter electrode

```
% LPR properties ――――――――――――  %
Aeff       = 0.04233;            % effective area of sensor    [cm^2]
                                 %
%% Tafel constants ―――――――――――  %
beta_a = 0.35;                   % anodic tafel constant       [V/dec]
beta_c = 0.15;                   % cathodic tafel constant     [V/dec]
                                 %
% Material properties for AL7075-T3 ―  %
AW         = (5.8*65.38+2.3*24.3050+...  % Atomic weight       [g/mol]
             1.4*63.546+90.5*26.9815386)/100;  %
z          = 3;                  % Electron loss per atom      [-]
EW         = AW/z;               % equivalent weight           [g/mol]
rho        = 2.810;              % density                     [g/cm^3]
                                 %
% Other constants ――――――――――――  %
F          = 9.6485e4;           % Faraday's constant          [C/mol]
                                 %
% Coupon Geometry ――――――――――――  %
l          = 7.8;                % Coupon length               [cm]
w          = 1.0;                % Coupon witdh                [cm]
SA         = 2*(l*w);            % Coupon surface area         [cm^2]
                                 %
%% Computation of Corrosion Rate ――――  %
B          = (beta_a*beta_c)/... % Proportionality constant    [V/dec]
             (2.303*(beta_a+beta_c));  %
Bcorr      = EW*B/(F*Aeff);      % Corrosion rate coefficient  [Ohm*(g/cm^2)/s]
```

Fig. 6.32 Sensor modeling parameters (*Note* The effective area of the sensor, Aeff, was found empirically. Based on the experimental results, the area parallel to the electrode pair (and not the top area of the sensor) was determined to be the effective area. Using the geometry of the sensor, Aeff was computed to be 0.04233 cm^2, which agreed with the experimental results)

separated by a scrim cloth made from Teflon with a thickness of 4μm. One of the μLPR electrodes was used as the working electrode while the other was used as the reference electrode. A semi-log plot of the applied electric potential versus the applied current is shown in Fig. 6.33. Superimposed on the plot are linear fits at the anode and cathode regions. The slopes of these lines were used to compute β_a and β_c as approximately 150 mV/dec and 350 mV/dec, respectively.

6.31 Results

The steps outlined in the procedure were repeated four times to arrive at four sets of mass loss measurements. A plot of the estimated and measured mass loss density versus time for each of the eight sensors and eight coupons is provided in Fig. 6.34. A scatter plot showing the comparison of the average measured mass loss versus the average estimated mass loss is given in Fig. 6.35. Towards the end of the experiment, sensors #3, #4 and #5 began to deviate from the average measured mass loss. This is also apparent from the time series plot of estimated corrosion rate for each of the eight μLPR sensors shown in Fig. 6.36. According to the plot, the corrosion rates for sensor #4 began to decrease, followed by sensor #5 and then sensors #3

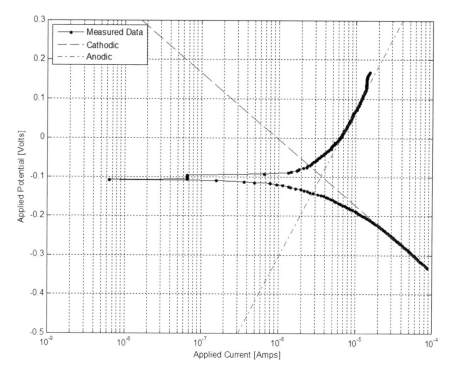

Fig. 6.33 Tafel plots for the μLPR sensor in a B117 solution (pH of 5.5)

and #6. Since each of the eight μLPR sensors are placed next to each other in ascending order, it is believed sensors #3–6 suffered from a localized disturbance in the flow of solution.

6.32 Model On-Line Update [24]

Important elements in the corrosion modeling include a time-varying parameter β and noises w and v. The parameter β describes the fault growth according to system operating conditions while noises w and v, to a certain extent, describe the confidence on the model. If a good model is developed, they can be selected each as a very small value. On the other hand, if a very rough model is used, they need to be selected as a large value. The trade-off is that when a large noise model is used, the estimated results tend to be noisy too. We will exploit uncertainty management tools developed at Georgia Tech to address this problem. We focus on the development of the model, or β. Noise models w and v are determined via data.

The parameter β depends on current environmental conditions affecting corrosion growth, i.e. relative humidity, wetness, pH, etc. the loading profile that is being

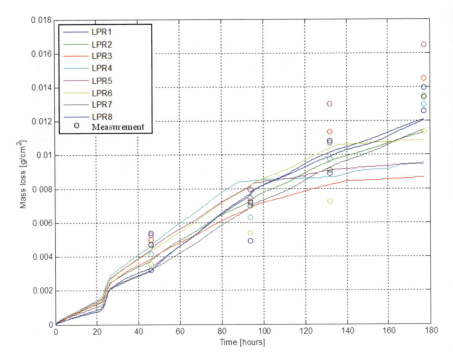

Fig. 6.34 Plot of estimated and measured mass loss density versus time for each of the eight sensors and eight coupons, respectively

applied to the component of interest. In prediction models, a population equation, like Paris' law, is adopted.

When the fault mode is corrosion, the fault dimension is often measured by the area of corrosion. Paris' Law as shown in (A5) characterizes the corrosion growth.

$$\dot{D} = \frac{dD}{dt} = C_D(D)^n \qquad (6.2)$$

Here, C_D and n are determined by an online adaptation routine. Based on Equation (A5), a defect or corrosion growth model can be written as

$$D(t + \Delta t) = D(t) + \Delta t C_D(D(t))^n \qquad (6.3)$$

To achieve the goal of parameter adjustment, an adaptive prediction scheme for the corroding panel is given in Fig. 6.37.

In this model, feature $F(t)$ is extracted from the collected mass loss or imaging time varying data $x(t)$. This information, combined with ground truth data about the defect area, is used to build a nonlinear mapping between the defect area and the feature. From this nonlinear mapping, the defect area at current time instant, $D(t)$, can be estimated. Additionally, the estimated defect area $D(t)$ is traced back to time

6 Corrosion Modeling

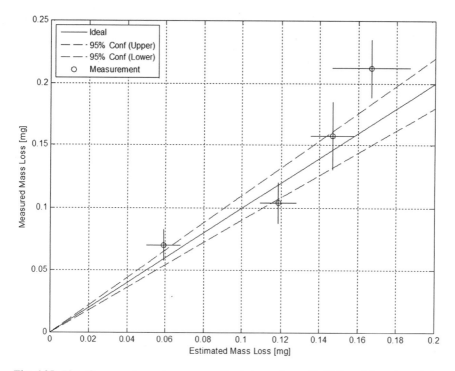

Fig. 6.35 Plot of measured mass loss versus estimated mass loss with 95% confidence boundaries

instant $t - \Delta$, resulting in $D(t - \Delta)$. This defect area $D(t - \Delta)$ and the operating conditions serve as the input to the prognostic model to predict the defect area at time instant t, denoted as $D_p(t)$. Then, $D(t)$ and $D_p(t)$ are compared to compute an error $e(t)$. Optimization methods can be introduced to adjust the parameters of the model to minimize $e(t)$. Note that in this method, there are two preconditions: ground truth data are available to build a nonlinear mapping between feature and defect area and an optimization routine.

Note that the propagation of the defect area under tightly controlled conditions could show significantly different behaviors. Therefore, the previous deterministic model must be modified to take into consideration this situation. Theoretically, the uncertainty is due to the stochastic characteristics of the progression equation and, therefore, it is reasonable to add a random variable into this formulation. In practice, adding a random variable is the same as adding a random variable into its parameters and we arrive at:

$$D(t + \Delta t) = D(t) + \Delta t C_D (D(t))^n$$
$$C_D = C_D + w_c \quad (6.4)$$
$$n = n + w_n$$

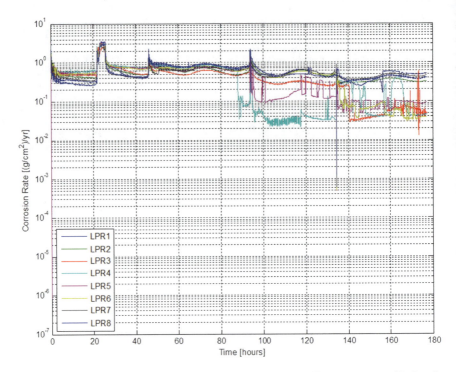

Fig. 6.36 Plot of estimated mass loss versus time using the μLPR measurements (During the hours 24 and 28 the pH of the solution was changed to 4. The polarization resistance reduced to near 2000 Ω. Since the lower limit of the AN110 hardware is 1000 Ω, the pH was returned to a 5), (Approximately every 48 h the sensors were removed and the coupons were cleaned. Initially and upon re-entering the solution, a spike in the corrosion rate was observed. This is visible in hours 0, 45, 92 and 136)

where C_D and n can be regarded as states associated with the model, w_c and w_n are zero mean random noise.

With unit step size, Eq. (6.4) can be modified as

$$D(t+1) = D(t) + p_1(t) C_D (D(t))^{p_2(t)n}$$
$$C_D = C_D + w_c \qquad (6.5)$$
$$n = n + w_n$$

Thus, two parameters $p_1(t)$ and $p_2(t)$ are introduced to facilitate the online parameter adaptation scheme.

To determine the parameters, a recursive least square algorithm with a forgetting factor is employed since it is generally fast in its convergence. The algorithm is implemented as follows:

6 Corrosion Modeling

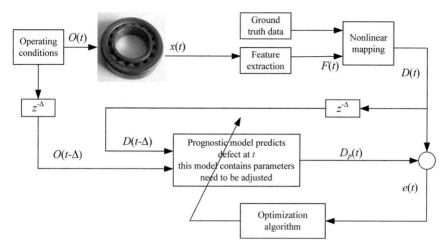

Fig. 6.37 The on-line prognostic adaptation model

Step 1: define a cost function as:

$$J(\theta) = \frac{1}{2}\sum_{t=1}^{T} \lambda^{T-t}\left[D(t) - D\left(\hat{\theta}(t-1)\right)\right]^2 \quad (6.6)$$

where λ is the forgetting factor, which is usually given in the range of $0 < \lambda \leq 1$, and $\theta = [p_1(t)p_2(t)]^T$ is the parameter vector to be determined.

Step 2: Calculate the derivatives with respect to parameters Θ:

$$\phi(t) = \frac{d\hat{D}(t,\theta)}{d\theta} \quad (6.7)$$

Step 3: The parameter update is given by:

$$\hat{\theta}(t) = \hat{\theta}(t-1) + P(t)\phi(t)\left[D(t) - D\left(\hat{\theta}(t-1)\right)\right] \quad (6.8)$$

and $P(t)$ is updated as

$$P(t) = \frac{P(t-1)}{\lambda}\left[1 - \frac{\phi(t)\phi^T(t)P(t-1)}{\lambda + \phi^T(t)P(t-1)\phi(t)}\right] \quad (6.9)$$

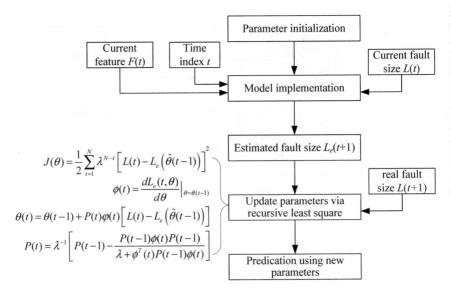

Fig. 6.38 Flow chart of the model on-line update

The recursive least square with a forgetting factor actually applies an exponential weighting to the past data. In the cost function (6.6), the influence of past data reduces gradually as new data become available. This algorithm can be easily applied on-line.

To implement the algorithm, a set of initial parameters must be given. Parameter $\theta(0)$ is given according to our prior knowledge of the system while $P(0)$ is given as a large number times an identity matrix.

A flow chart of this on-line adaptation algorithm is given in Fig. 6.38. The first step is to initialize the parameters used in the model. Currently, the initial value is obtained via a priori knowledge of the system. With more experimental data, some learning algorithms, such as neural networks, can be implemented to train the model so that its initial value is close to the actual one. The on-line adaptation routine will be more efficient with good initial parameters.

Note that the previous parameter adaptation is realized by a recursive least square method. Some other methods, such as an extended Kalman filter or a neural network, etc., can be used as well.

6.33 Consideration of Operating Conditions

In the previous model, the operating conditions such as ambient temperature, humidity, pH, etc., are not taken into consideration. The influence of the operating and environmental conditions is reflected by the on-line parameter tuning.

6 Corrosion Modeling

Moreover, stress conditions must be accounted for when appropriate. If the operating conditions can be compensated, a precise fault propagation model can be derived.

Let us consider the corrosion fault mode. It is known that humidity is a leading contributing factor of corrosion. Therefore, the environmental humidity should be incorporated in the corrosion propagation model. Suppose that the nominal humidity (could be normal room humidity) is denoted by H_n. The environmental humidity is measured as H_c. The normalized humidity condition then can be described as a humidity index h_i which is given as $h_i = H_c/H_n$. Clearly, large humidity values result in larger h_i, while small humidity values result in smaller h_i.

If we know that the humidity influences linearly the corrosion propagation, the previous Eq. (6.5) can be further re-written as Eq. (6.10) to include the humidity factor.

$$\dot{D} = \frac{dD}{dt} = h_i C_D (D)^n \tag{6.10}$$

If, however, we know that humidity influences exponentially the corrosion propagation, Eq. (6.5) can be re-written as:

$$\dot{D} = \frac{dD}{dt} = C_D (D)^{nh_i} \tag{6.11}$$

Accordingly, the discrete time form model should be modified as:

$$D(t+1) = D(t) + h_i C_D (D(t))^n \tag{6.12}$$

and

$$D(t+1) = D(t) + C_D (D(t))^{nh_i}, \tag{6.13}$$

respectively.

Relative humidity may also be included as a time-varying parameter in the model, just as other model parameters. When other operating conditions are included, such as nominal temperature T_n (normal operating temperature under healthy conditions), can be defined as well. Then, a temperature index t_i is assigned to represent this condition. Multiple ways are available to combine them into a single model.

Suppose these factors influence linearly the fault propagation, a possible alternative is to write the model as either

$$D(t+1) = D(t) + h_i t_i C_D (D(t))^n \tag{6.14}$$

or

$$D(t+1) = D(t) + (w_h h_i + w_t t_i) C_D (D(t))^n \qquad (6.15)$$

where w_h and w_t are weighting factors and $w_h + w_t = 1$.

The model can be written as either

$$D(t+1) = D(t) + C_D(D(t))^{nh_i t_i} \qquad (6.16)$$

or

$$D(t+1) = D(t) + C_D(D(t))^{n(w_h h_i + w_t t_i)} \qquad (6.17)$$

when the factors are exhibiting an exponential dependence.

It is possible that the influence of some factors is exhibiting a linear behavior, while that of others is exponential. In this case, suppose the linear factors are $f_{l,1}$ and $f_{l,2}$ and the exponential factors are $f_{e,1}$ and $f_{e,2}$. Then, the model can be written as

$$D(t+1) = D(t) + f_{l,1} f_{l,2} C_D(D(t))^{nf_{e,1} f_{e,2}} \qquad (6.18)$$

or

$$D(t+1) = D(t) + (w_{l,1} f_{l,1} + w_{l,2} f_{l,2}) C_D(D(t))^{n(w_{e,1} f_{e,1} + w_{e,2} f_{e,2})} \qquad (6.19)$$

where $w_{l,1}, w_{l,2}, w_{e,1}, w_{e,2}$ are weighting factors and $w_{l,1} + w_{l,2} = 1$ and $w_{e,1} + w_{e,2} = 1$.

6.34 Other Failure Prediction Models

Alternative model structures are available and may be appropriate under particular fault model conditions. Typical examples in this category include:

Kotzalas and Harris [25] suggested the following model:

$$\frac{da}{dN} = C(W_a)^m$$

where $W_a = (\sigma_{max} + \tau_{avg})\sqrt{\pi a}$ with σ_{max} being the maximum stress and τ_{avg} the average shear stress. Writing this in discrete form, we have

$$a_{k+1} = a_k + C(\sqrt{a_k})^m$$

Zhang et al. [26] suggested that, for a given bearing, the impulse magnitude is proportional to the bearing rotational speed and the spall size

$$F\Delta t = \frac{m\pi d_m n(1-\gamma)^2}{60D} x$$

where F is the mean impact force (N), Δt the impact duration (sec), D the roller diameter (mm), m mass of the roller (kg), d_m the pitch diameter of the bearing (mm), n shaft rotating speed (rpm), γ the ratio between D and d_m, and x the spall size (mm).

Choi and Liu [27] provide the following crack propagation model based on Paris' law:

$$\frac{da}{dN} = C \cdot \frac{H_b}{H_1} (\Delta K)^n$$

where H_b and H_1 are the Knoop hardness number at the bulk material sand local Knoop hardness number, respectively. The reason to include hardness in the model is because it affects the crack propagation and is shown experimentally that the crack propagates relatively faster as the local hardness is low if the stress field is identical [28].

Since the stress field over a whole crack is variable, ΔK in Paris' law is not a constant and should be compensated. To this end, the range of ΔK is calculated as

$$\Delta K = K_{max} - K_{min}$$

with K_{max} and K_{min} the maximum and minimum stress at the leading tip, respectively. To determine K_{max} and K_{min}, the following equation is used:

$$K_a = \sqrt{\frac{2}{\pi a}} \int_0^a \tau_c(x_a - \zeta) \left\{ \frac{a-\zeta}{\zeta} \right\}^{1/2} d\zeta$$

where a is the crack length, τ_c is the net shear stress, and x_a is the position of the leading tip.

Ioannides and Harris [29] include the effect of fatigue limit stress in the life prediction

$$\ln\left(\frac{1}{\Delta S_i}\right) \propto \frac{N^e (\sigma_i - \sigma_u)^c \Delta V_i}{z_i^h}$$

where

S_i = probability of survival of ΔV_i
N = number of stress cycles endured
Z = distance below surface to ΔV_i
e = Weibull slope or shape parameter

c = stress exponent parameter to ΔV_i
σ_i = stress at ΔV_i (sigma subi)
σ_u = fatigue limit stress (sigma subu)
h = experimentally determined parameter.

Qiu et al. [30] considered the bearing system as a single degree of freedom vibration system. The natural frequency and its amplitude at the natural frequency are related to the system stiffness. Since the relationship between failure lifetime, running time and stiffness can be established from the damage mechanics, the natural frequency and the acceleration amplitude of a bearing system can be related to its failure time.

The relationship is given as

$$\left(\frac{S_0}{S_d}\right)^\gamma = \eta\left(1 - \left(\frac{N}{N_{fe}}\right)^q\right)$$

where S_0 and S_d denote the amplitude of natural frequency under initial conditions and damage conditions, respectively. $\gamma = 2, q, \eta$, are coefficients depending on the operating conditions, materials and structure of the system, N is running cycles and N_{fe} is the estimated failure lifetime. The estimated lifetime in the model, and parameters like η can be determined via estimation algorithms. **He and Bechhoefer** [31] extended this model by using condition indicators rather than the amplitude of the natural frequency.

$$\left(\frac{CI_0}{CI_d}\right)^\gamma = \eta\left(1 - \left(\frac{N}{N_{fe}}\right)^q\right)$$

where CI_0 and CI_d denote the condition indicators under normal and damage conditions, respectively.

6.35 A Corrosion Modeling Framework for Steel Structures

A Failure Modes and Effects Analysis (FMEA) was performed for stainless steel parts to identify credible degradation mechanisms that may be active during the lifespan of the steel components.

Pitting occurs at locations where the passive surface film breaks down and allows the dissolution of the underlying metal. Despite the often-slow rate of pit penetration, pits can serve as initiation sites for through wall cracking, which progresses much more rapidly.

Corrosion cracking (NUREG/CR-7116, SRNL-STI-2011-00005, "Materials Aging Issues and Aging Management for Extended Storage and Transportation of Spent Nuclear Fuel," November 2011. (Available with NRC Accession No. ML11321A182).

The environmental factors that increase the cracking susceptibility include higher temperatures, increased chloride content, lower pH, and higher levels of tensile stress. Temperature is an important variable.

6.36 Modeling of Nuclear Waste Storage Facilities

It has been established that corrosion is one of the most important factors causing deterioration, loss of metal, and ultimately decrease of nuclear waste management facilities performance and reliability in such critical systems. Corrosion monitoring, data mining, accurate detection and quantification are recognized as key enabling technologies to reduce the impact of corrosion on the integrity of these assets. Accurate and reliable detection of corrosion initiation and propagation with specified false alarm rates requires novel tools and methods. Corrosion states take various forms starting with microstructure corrosion and ending with stress induced cracking [32–36].

Figure 6.39 depicts a typical view of canister where waste fuel is stored.

Fig. 6.39 Typical canister for nuclear waste fuel storage

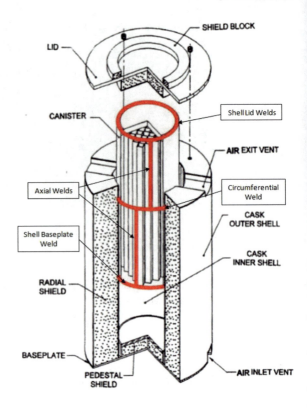

6.37 State of the Art in Corrosion Modeling for Nuclear Storage Facilities

Figure 6.40 shows a framework for nuclear waste storage sensing, monitoring and modeling. A number of studies have been carried out by government, national laboratories and industry to characterize the corrosion processes in nuclear waste fuel canisters [37–42]. We summarize a few findings.

DOE performed polarization-resistance measurements at 60, 80, and 100 °C in solutions made by dissolving various amounts of sodium chloride and potassium nitrate to obtain solutions ranging between 1 and 6 molal chloride and 0.05 and 3.0 molal nitrate. A total of 360 polarization-resistance measurements were made, and corrosion rates were calculated from the measurements [24].

The corrosion rates then were fit to an Arrhenius-type relationship,

$$RT = A \exp(Ea/RT)$$

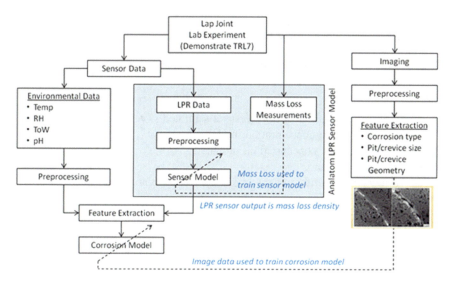

Fig. 6.40 Framework for nuclear waste storage facilities modeling

where

RT is the generalized corrosion rate, nm/yr,
A is the pre-exponential factor, nm/yr,
Ea is the apparent activation energy, J/mol,
R is the gas constant, 8.314 J/mol K, and
T is temperature, K.

The resultant fit yielded an apparent activation energy of 40.78 kJ/mol with a standard deviation of ±11.75 kJ/mol.

6.38 Localized Corrosion

For predicting the postclosure performance of the Alloy 22 outer barrier of the waste package with regard to localized corrosion, DOE divides localized corrosion into two parts: localized corrosion caused by seepage of water onto the waste package surface, and localized corrosion due to deliquescence of hydroscopic salts in dust deposited on the waste package by ventilation in the 50–100 year (or longer) period before closure of the repository.

To determine Ecorr, DOE has measured open-circuit potentials versus time for various Alloy 22 samples in model aqueous solutions over a temperature range of 25–90 °C for periods of up to 3 years. DOE then uses multiple linear regression on the data to fit a model for Ecorr as a function of nitrate concentration, chloride concentration, temperature, and pH, i.e.:

$$\text{Ecorr} = f([NO_3], [Cl], T, pH)$$

where

Ecorr is corrosion potential in mV versus a saturated silver chloride electrode,
$[NO_3]$ is the nitrate ion concentration, molal
$[Cl]$ is the chloride ion concentration, molal,
T is temperature, °C, and
pH is the negative logarithm of the hydrogen ion concentration.

To determine Ecrit, DOE uses the ASTM G61-86 cyclic potentiodynamic polarization (CPP) technique to determine the repassivation potentials [17] for a number of Alloy 22 samples in a variety of aqueous solutions. The crossover point, ER, is where the forward scan intersects the reverse scan. The resultant data were fit to express critical potential as a function of nitrate concentration, chloride concentration, and temperature, i.e.:

$$\text{Ecrit} = g([NO_3], [Cl], T)$$

where Ecrit is critical potential in mV versus a saturated silver chloride electrode, and $[NO_3]$, $[Cl]$, and T are as defined above.

More information about the data and the function, g, may be found in Ref. [43].

Comments DOE assumes that localized corrosion of Alloy 22 will initiate if Ecorr, as represented by function f, is greater than Ecrit, as represented by function g. The data on which function f is based are long-term data—up to 3 years in many cases but at least 8 months in all cases. On the other hand, the data on which function g is based are short-term data—a matter of a few days. The question of whether mixing short-term data and long-term data to form the basis for predicting localized corrosion initiation is appropriate then must be asked. If so, is doing so likely to overpredict the occurrence of localized corrosion or under predict it? In addition, despite the long-term nature of the corrosion potential experiments, it is not clear that all the experiments had reached a stable value by the end of the tests, and some of the test results were noisy at the end of the tests, also making the stability of the test results questionable.

A problem with interpreting some of the data is that DOE's tests of long-term corrosion, many with crevice formers and others boldly exposed, apparently do not corroborate the model that DOE has developed by fitting Ecorr and Ecrit equations to the electrochemical data. For example, according to the localized corrosion model, some of the tests used to develop the generalized corrosion model discussed above should have shown localized corrosion. However, none of them did. Similarly, many other long-term corrosion tests that the model indicates should have developed localized corrosion did not. With a single exception, the only tests where the model predicted localized corrosion and localized corrosion occurred were in highly concentrated solutions of calcium chloride with minor amounts (or none) of calcium nitrate. Such solutions are not very likely to exist in the repository. In summary, the model used to predict localized corrosion under seepage conditions appears to be overly conservative in that it predicts that localized corrosion will occur in many instances where experimental data indicate that it will not occur—at least not in the time frame of the experiments.

Propagation Once localized corrosion initiates, the assumption is that it will propagate at a constant rate, ranging from 12.7 to 1270 µm/yr. Apparently, these values were selected from Alloy-22 uniform corrosion rates in highly aggressive solutions extracted from the product literature of one of the manufacturers of Alloy 22. Such data should be used only to compare the relative corrosion resistance of different alloys, and localized corrosion rates, in general, are much higher than uniform corrosion rates. However, using the highest published corrosion rates above results in penetration of the waste packages, once localized corrosion has been initiated, in a matter of 20 years, which, for a million-year (or longer) period of concern, is essentially instantaneously.

It is not clear from currently available documentation how DOE models what happens after the waste package is penetrated by localized corrosion, i.e., what the area, morphology, and geometry of the penetration(s) are. One approach could be to assume that the entire area contacted by seepage disappears when penetration occurs. This would be an extremely conservative view and inconsistent with the very nature of localized corrosion. Even more extreme would be to assume that the entire waste package disappears at the time of penetration. In either case, the corrosion resistance of the container alloys essentially becomes irrelevant, and the containment of dangerous radionuclides becomes the responsibility of the waste form and natural barriers.

6.39 Localized Corrosion Due to Deliquescence

Brines can form on waste package surfaces at temperatures of up to 210 °C because of deliquescence, so the possibility of localized corrosion during the thermal pulse period (Essentially the ∼100- to 1000-year period immediately after repository closure, during which the waste package surfaces are above boiling) is a concern. Localized (crevice) corrosion has been observed in autoclave (pressurized) experiments performed on Alloy 22 in aqueous solutions of 2.5 m and 6.4 m Cl^- with $[NO_3^-]/[Cl^-]$ ratios of 0.5 or 7.4 at temperatures of 160 and 220 °C. 6 Localized corrosion was observed in all cases but was not anticipated in the solutions with nitrate-to-chloride ratios of 7.4. The test solutions were made by dissolving sodium chloride, sodium nitrate, and potassium nitrate. Brines containing sodium, potassium, chloride, and nitrate cannot form at atmospheric pressure in the higher end of the temperature range of interest unless the nitrate-to-chloride ratio of the brine is well above 1. However, if the nitrate concentrations are lower than anticipated or if the nitrate is removed by physical or biological processes, whether the stable brine would decrease in amount, form a metastable brine, or solidify (thus being rendered innocuous) has not been determined. In addition, whether other brines could form at high temperatures from other mixtures of salts that may exist at Yucca Mountain has not been explored systematically. In other words, the possibility of other high-temperature corrosive environments has not been ruled out. The test solutions were made by dissolving sodium chloride, sodium nitrate, and potassium nitrate. Brines containing sodium, potassium, chloride, and nitrate cannot form at atmospheric pressure in the higher end of the temperature range of interest unless the nitrate-to-chloride ratio of the brine is well above 1. However, if the nitrate concentrations are lower than anticipated or if the nitrate is removed by physical or biological processes, whether the stable brine would decrease in amount, form a metastable brine, or solidify (thus being rendered innocuous) has not been determined. In addition, whether other brines could form at high temperatures from other mixtures of salts that may exist at Yucca Mountain has not been explored systematically. In other words, the possibility of other high-temperature corrosive environments has not been ruled out.

The Board sponsored a workshop on localized corrosion in September 2006 to discuss deliquescence-based corrosion of Alloy 22 in repository-relevant environments. At that meeting, DOE presented a strong case suggesting that the nitrate-to-chloride ratios in the repository were sufficiently high that localized corrosion could not occur. In addition, several workshop participants suggested that the propagation of localized corrosion would effectively be stifled because (a) migration rates for nitrate into occluded regions are higher than chloride migration rates and repassivation would occur within the occluded regions and/or (b) the amount of water and/or aggressive species is so low that the occluded regions would effectively be "starved" as the damage propagated and localized corrosion would essentially halt. The Board made two recommendations to DOE as a result of the workshop: (1) determine the level of nitrate needed to inhibit localized corrosion over the *entire* temperature range (i.e., up to 210 °C), and (2) determine the relative migration rates for the migration of nitrate and chloride ions into crevices. At this time, neither of these recommendations appears to have been implemented, and thus the possibility of localized corrosion due to deliquescence at high temperatures remains uncertain.

At lower temperatures, i.e., waste package surface temperatures of 100 °C and below, many pure salts and salt mixtures will persist on waste package surfaces that can deliquesce in the near-100% relative humidity environment that will exist in the repository after the thermal pulse. Some of these salts can form corrosive brines, according to the Ecorr, Ecrit localized corrosion model discussed above.

6.40 Epilogue

A large variety of corrosion modeling tools/methods have been introduced over the past decades addressing multiple corrosion stages and processes. We addressed in this chapter the fundamental concepts and the modeling methods for corrosion initiation as an electrochemical process to analytical, mostly semi-empirical, tools that represent a large variety materials and processes affected by corrosion. They constitute the cornerstone for accurate diagnosis and prognosis. They may be used for data generation and verification/validation of corrosion remediation strategies. It is obvious that the lack of "good" data relating to corrosion processes is hindering the development and utility of modeling techniques. Such models to be useful in more general terms, the data employed for their development and validation must be correlated to actual environmental conditions, time-stamped over a long-time interval and statistically sufficient. This need is well understood by government and industry and means/programs are being instituted to address this crucial issue.

References

1. Wallace W, Hoeppner DW (1985) AGARD corrosion handbook volume I aircraft corrosion: causes and case histories. AGARD-AG-278, vol 1
2. Wei RP, Liao CM, Gao M (1998) A transmission electron microscopy study of 7075-T6 and 2024-T3 aluminum alloys. Metall Mater Trans A 29A:1153–1163
3. Hoeppner DW, Chandrasekaran V, Taylor AMH (1999) Review of pitting corrosion fatigue models. In: International committee on aeronautical fatigue, Bellevue, WA, USA
4. Huang T-S, Frankel GS (2006) Influence of grain structure on anisotropic localized corrosion kinetics of AA7xxx-T6 alloys. Corros Eng Sci Technol 41(3):192–199
5. McAdam G, Newman PJ, McKenzie I, Davis C, Hinton BR (2005) Fiber optic sensors for detection of corrosion within aircraft. Struct Health Monit 4:47–56
6. Pidaparti RM (2007) Strucural corrosion health assessment using computational intelligence methods. Struct Health Monit 6(3):245–259
7. Rao KS, Rao KP (2004) Pitting corrosion of heat-treatable aluminum alloys and welds: a review. Trans Indian Inst Met 57(6):593–610
8. Frankel GS (1998) Pitting corrosion of metals: a review of the critical factors. J Electrochem Soc 145(6):2186–2198
9. Szklarska-Smialowska Z (1999) Pitting corrosion of aluminum. Cossorion Sci 41(9):1743–1767
10. Pereira MC, Silva JW, Acciari HA, Codaro EN, Hein LR (2012) Morphology characterization and kinetics evaluation of pitting corrosion of commercially pure aluminum by digital image analysis. Mater Sci Appl 3(5):287–293
11. Forsyth DS, Komorwoski JP (2000) The role of data fusion in NDE for aging aircraft. SPIE Aging Aircr Airpt Aerospe Hardw IV 3994:6
12. Brown D, Darr D, Morse J, Laskowski B (2012) Real-time corrosion monitoring of aircraft structures with prognostic applications. In: Annual conference of the prognostics and health management society, vol 3
13. Brown DW, Connolly RJ, Laskowski B, Garvan M, Li H, Agarwala VS, Vachtsevanos G (2014) A novel linear polarization resistance corrosion sensing methodology for aircraft structure. In: Annual conference of the prognostics and health management society, vol 5, no 33

14. Kawai S, Kasai K (1985) Considerations of allowable stress of corrosion fatigue (focused on the influence of pitting). Fatigue Fract Eng Mater Struct 8(2):115–127
15. Belanger P, Cawley P, Simonetti F (2010) Guided wave diffraction tomography within the born approximation. IEEE Trans UFFC 57:1405–1418
16. Provan JW, Rodriguez ES (1989) Part I: Development of a markov description of pitting corrosion. Corrosion 45(3):178–192
17. Bolzoni F, Fassina P, Fumagalli G, Lazzari L, Mazzola E (2006) Application of probabilistic models to localised corrosion study. Metallurgia Ital 98(6):9–15
18. Timashev SA, Malyukova MG, Poluian LV, Bushiskaya AV (2008) Markov description of corrosion defects growth and its application to reliability based inspection and maintenance of pipelines. In proceedings of the 7th ASME International Pipeline Conference (IPC '08), Calgary, Canada, Sep 2008
19. López De La Cruz J, Lindelauf RHA, Koene L, Gutiérrez MA (2007) Stochastic approach to the spatial analysis of pitting corrosion and pit interaction. Electrochem Commun 9(2):325–330
20. Harlow DG, Wei RP (1994) Probability approach for prediction of corrosion and corrosion fatigue life. AIAA J 32(10)
21. Hoeppner DW, Goss GL (1971) A new apparatus for studying fretting fatigue. Rev Sci Instrum 42:817
22. Lindley TC, Mcintyre P, Trant PJ (1982) Fatigue-crack initiation at corrosion pits. Metals Technol 9(1):135–142
23. Kondo Y (1989) Prediction of fatigue crack initiation life based on pit growth. Corrosion 45(1):7–11
24. Li H, Garvan M, Li J, Echauz J, Brown D, Vachtsevanos G, Zahiri F (2017) An integrated architecture for corrosion monitoring and testing, data mining, modeling, and diagnostics/prognostics. Intl J Progn Health Manage 8(5):12
25. Kotzalas MN, Harris TA (2001) Fatigue failure progression in ball bearings. J Tribol 123:238–242
26. Zhang C, Kurfess T, Danyluk S, Liang S (1999) Dynamic modeling of vibration signals for bearing condition monitoring. The 2nd international workshop on structural health monitoring, Stanford, pp 926–935
27. Choi Y, Liu C (2006) Rolling contact fatigue life of finish hard machined surfaces Part I. Model develop Wear 261:485–491
28. Davies M, Chou Y, Evans C (1996) On chip morphology, tool wear and cutting mechanics in finish hard turning. Ann CIRP 45(1):77–82
29. Ioannides E, Harris T (1985) A new fatigue life model for rolling bearing. Trans ASME J Tribol 107:367–278
30. Qiu J, Zhang C, Seth B, Liang S (2002) Damage mechanics approach for bearing lifetime prognostics. Mech Syst Signal Process 16(5):817–829
31. He D, Bechhoefer E (2008) Bearing prognostics using HUMS condition indictors. American Helicopter Society 64th Annual Forum, Montreal, Canada
32. United States Nuclear Regulatory Commission (2009) Standard review plan for spent fuel dry storage systems at a General License Facility
33. NRC SFST ISG-11 (2003) Revision 3, Division of Spent Fuel Storage and Transportation Interim Staff Guidance—1, Revision 2, Classifying the condition of spent nuclear fuel for interim storage and transportation based on function
34. United States Nuclear Waste TECHNICAL REVIEW BOARD (2010) Evaluation of the technical basis for extended dry storage and transportation of used nuclear fuel
35. Grossbeck ML (Faculty Advisor), ICaMS (Internal Cask Monitoring System) (2008) A conceptual report on the development of the information from the interior of a spent fuel cask, undergraduate entry, American Nuclear Society Student Design Competition, Nuclear Engineering Department, The University of Tennessee Knoxville
36. Grossbeck ML (Faculty Advisor) (2008) Device to transmit critical information from the interior of a spent fuel cask, undergraduate entry. American Nuclear Society Student Design Competition, Nuclear Engineering Department, The University of Tennessee Knoxville

37. NAC-MPC final safety analysis report, Docket number 72-1025
38. NAC-UMS final safety analysis report, Docket number 72-1015
39. MAGNASTOR final safety analysis report, Docket number 72-1031
40. Liu YY (2012) Monitoring, INMM spent fuel seminar, Washington, DC, Jan 31–Feb 2, 2012
41. Bakhtiari S, Wang K, Elmer TW, Koehl E, Raptis AC (2013) Development of a novel ultrasonic temperature probe for long-term monitoring of dry cask storage systems. AIP Conf Proc 1511:1526
42. Carstens TA, Corradini ML, Blanchard JP, Ma Z (2012) Thermoelectric powered wireless sensors for spent fuel monitoring. IEEE Trans Nucl Sci **59**(4)
43. A. S. G102 (2004) Standard practice for calculation of corrosion rates and related information from electrochemical measurements. ASTM International, West Conshohocken

Chapter 7
Corrosion Diagnostic and Prognostic Technologies

George Vachtsevanos

Abstract This chapter addresses the development and utility of fundamental diagnostic and prognostic algorithms to assist in early detection of corrosion initiation and progression so that immediate remediation can be taken to avoid further structural deterioration while limiting significantly repair and replacement costs. Corrosion, in its different stages, is a significant challenge affecting the operational integrity of a vast variety of equipment and processes. Corrosion prevention costs are amounting to billions of dollars each year. As complex equipment age, exposure to corrosion processes is increasing at a substantial and alarming rate contributing to equipment degradation and leading to failure modes. Major efforts have been underway over the past years to develop and implement corrosion prevention and protection materials/processes to extent the useful life of critical equipment/facilities preventing rapid deterioration and retirement. Early corrosion detection is urgently required to warn the operator/maintainer of impending detrimental events that endanger the integrity and life of critical aerospace and industrial processes exposed to corrosive environments. Accurate prediction of the growth of corrosion states is an essential component of the architecture aiming to provide estimates of the time remaining for remediation while the system/process is required to complete a current task or mission. The enabling technologies build upon the sensing

G. Vachtsevanos (✉)
Georgia Tech, Atlanta, USA
e-mail: gjv@ece.gatech.edu

© Springer Nature Switzerland AG 2020
G. Vachtsevanos et al. (eds.), *Corrosion Processes*, Structural Integrity 13, https://doi.org/10.1007/978-3-030-32831-3_7

modalities, corrosion modeling tools and methods detailed in previous chapters. Corrosion modeling has been addressed over the past years from multiple investigators on behalf of government agencies and industry (see Chapter on Corrosion Modeling). We take advantage of these efforts to formulate the corrosion diagnostic and prognostic algorithms. We use case studies and examples illustrating the theoretical developments.

Keywords Corrosion detection · Corrosion prediction · Corrosion assessment · Diagnostics performance metrics · Prognostics performance metrics

7.1 The Corrosion Detection and Prediction Architecture

Assessing the potential impact of corrosion processes on the integrity of critical military and industrial systems, aircraft, transportation and industrial processes, requires new and innovative technologies that integrate robust corrosion monitoring, data mining, corrosion detection and prediction of the corrosion (pits, crevices, cracks) growth rate with intelligent reasoning paradigms that capture historical data, expert opinions and adaptation strategies to associate current evidence with past cases obtained fleet-wide for similar system components. Figure 7.1 depicts the modules of the overall framework from sensing to data mining, corrosion modeling, diagnostic and prognostic strategies to means for corrosion mitigation. We are proposing a holistic framework to assess the impact of corrosion-induced processes, on typical aluminum alloy components and other metals that begins with methods/tools for on-platform sensing, data processing, corrosion modeling of all corrosion stages of particular interest in this study. These functions support diagnostic and prognostic algorithms that are designed to meet customer requirements/specifications for confidence/accuracy and false alarm rates while managing effectively large-grain uncertainty prevalent in health management studies of engineering systems. The hardware/software components of the sensing and health management system form a "smart" sensor that monitors, processes data/images and decides on-line in real time on the health status and future progression of corrosion pitting/cracking. Corrosion monitoring, detection and prediction entail a series of functions. Starting with the monitoring apparatus, data/image collection and processing, corrosion modeling, detection and prediction and, finally, assessment of the potential impact of corrosion on the operational integrity of an asset. This sequence of events is shown schematically in Fig. 7.1. Corrosion states take various forms starting with microstructure corrosion and ending with stress induced cracking, as outlined in Fig. 7.2.

The sensing, modeling and diagnostic/prognostic functions are coupled with a novel reasoning paradigm, called Dynamic Case Based Reasoning (DCBR) that houses a case library composed of past documented cases detailing the impact of cracking on the integrity of platform components/systems. The DCBR is supported by cognitive routines for learning and adaptation so that new evidence is compared

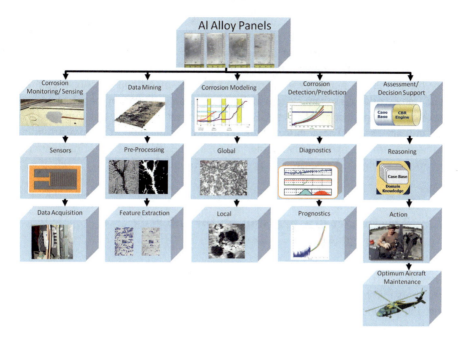

Fig. 7.1 A conceptual representation of the enabling technologies for corrosion assessment

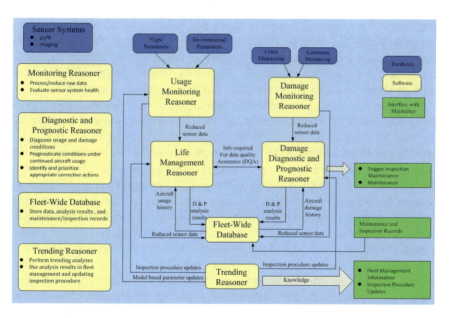

Fig. 7.2 A general architecture for an aircraft structural health management system

with stored cases and those occurring for the first time are "learned" by the reasoner. Figure 7.2 depicts the main modules of the proposed framework. The schematic represents a general architecture for an aircraft corrosion/crack monitoring, the reasoning modules employed to detect and predict the extent of cracking/corrosion, the life management component and the maintenance actions required. The framework stems from current and past research in corrosion modeling and the development/application of novel CBM+ practices introduced by this research team for military assets. The architecture is set as a decision support system providing advisories to the operator/maintainer as to the health status of critical aircraft component subjected to corrosion and in need of corrective action.

7.2 The Impact of Corrosion on the Integrity of Critical Assets

Over the past years, it has usually been the high fatality spectacular catastrophic accidents that have worked as the catalyst for change. Historical evidence suggests that fatigue due to corrosion cracking is a major contributor to aircraft accidents. Cracking of critical aircraft structures may endanger severely the performance and life of the vehicle. Corrosion damage can sometimes be greatly exaggerated by the circumstances. While many of the accidents due to failed corroded components have gone non-public for reasons of liability or simply because the evidence disappeared in the catastrophic event, others have made the headlines. The structural failure on April 28, 1988 of a 19-year-old Boeing 737, operated by Aloha airlines, was a defining event in creating awareness of aging aircraft in both the public domain and in the aviation community. Numerous other aircraft catastrophic events were attributed to corrosion accelerated fatigue as the failure mechanism.

Recent events have demonstrated the importance of early and accurate detection and prediction of the severity and impact of such corrosion-induced cracking and the need for immediate remediation/prevention to avoid catastrophic consequences or increased financial burden. In the recent past, cracks on a/c structures detected during regular maintenance have necessitated urgent actions to be taken to improve the design and installation of failing components. The pictures in Fig. 7.3 show the catastrophic effects of corrosion and cracking. Many of these incidents were attributed to corrosion/cracking fatigue. A systematic, thorough and robust corrosion modeling effort, addressing all corrosion stages for aluminum alloys or other metals, from micro to meso and macro levels, combined with appropriate sensing, data mining and decision support tools/methods (diagnostic and prognostic algorithms) may lead to substantially improved structural component (materials, coatings, etc.) performance and reduced exposure to detrimental consequences. Reliable, high-fidelity corrosion models form the foundation for accurate and robust corrosion detection and growth prediction. A suitable modeling framework assists in the development, testing and evaluation of detection and prediction algorithms.

7 Corrosion Diagnostic and Prognostic Technologies 235

Fig. 7.3 Catastrophic effects of corrosion and cracking

It may be employed to generate data for data-driven methods to diagnostics/prognostics, test and validate routines for data processing tool development, among others. The flexibility provided by a simulation platform, housing appropriate detection and progression models, is a unique attribute in the study of how corrosion processes are initiated, evolving and may be, eventually, mitigated in physical systems.

We define a **severity index** resulting from the application of verifiable data mining, diagnostic/prognostic algorithms in real time on-platform aimed to indicate when cracking must be attended to in order to extend the life of critical components, reduce the cost of corrosion prevention and avoid detrimental events. These developments are coupled with current research efforts aiming to design and implement on-platform a "smart" sensing modality that will perform all necessary functions from early detection to prognosis and estimation of the severity of such events. We will rely on a reasoning paradigm built from past historical evidence, learning and adaptation capabilities to assess the severity of cracking and assign an index to the current situation.

7.3 Corrosion Processes

Of particular interest to our theme is localized corrosion and cracking, i.e. cracking initiating at points on the surface of a specimen (joints, fasteners, bolts, etc.). A metal surface (aluminum alloy, etc.) exposed to a corrosive environment may, under certain conditions experience attack at a number of isolated sites. If the total area of these sites is much smaller than the surface area then the part is said to be experiencing localized corrosion. Figure 7.4 shows schematically the progression from pitting to cracking of corroding specimens. The rate of dissolution in this situation is often much greater than that associated with uniform corrosion and structural failure may occur after a very short period. Several different modes of

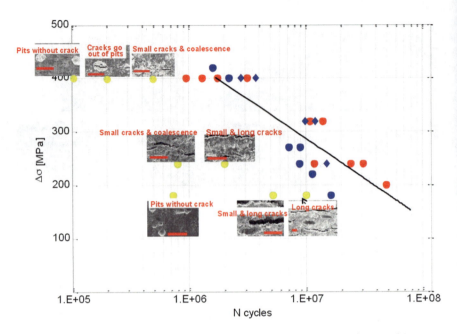

Fig. 7.4 From pitting to cracking of corroding specimens (*source* Dr. Vinod Agarwala)

localized corrosion may be identified. These are dependent on the type of specimen undergoing corrosion and its environment at the time of attack. Most destructive forms are pitting corrosion which is characterized by the presence of a number of small pits on the exposed metal surface, crevice attack and cracking. The rapidity with which localized corrosion can lead to the failure of a metal structure and the extreme unpredictability of the time and place of attack, has led to a great deal of study of this phenomenon. In this localized view, imaging studies are focusing on small areas of the global image where corrosion initiation is suspected and may spread more rapidly than other areas. We will exploit novel image processing tools/methods, in combination with other means (mass loss calculations) to identify features of interest to be used in the modeling task, since imaging of corroding surfaces offers a viable, robust and accurate means to assess the extent of localized corrosion.

Figure 7.5 shows a conceptual schematic of the major modules of the corrosion processing and detection/prediction architecture. The architecture combines a model-based and data-driven methodology taking advantage of experimental data, corrosion progression models, and an estimation method called particle filtering in order to detect the early initiation of corrosion. It is accompanied by performance metrics for detection confidence, false alarm rate, and prediction accuracy/precision [1]. In this architecture, the most important components supporting the implementation of the algorithms are feature extraction, fault/degradation diagnosis, and failure prognosis.

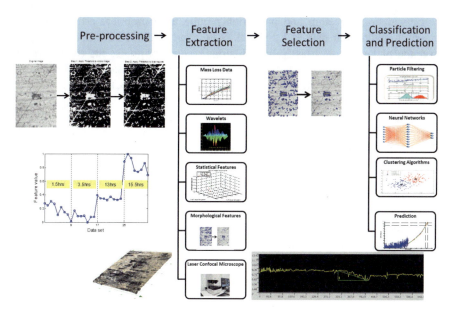

Fig. 7.5 The processing and detection/prediction architecture

7.4 Data and Corrosion Modeling Requirements

Advanced corrosion health assessment systems require comprehensive quantitative information, which can be categorized into a variety of feature groups, such as corrosion morphology, texture, and location. Implementation of advanced health assessment systems will require the exploration of new testing methods and data fusion methods from multiple testing techniques. Forsyth and Komorowski [2] discussed data fusion techniques to combine the information from multiple NDE techniques into an integrated form for structural modeling. Several other studies have looked into different sensing technologies for corrosion health monitoring, including the use of a micro-linear polarization resistance (μLPR) sensor [3, 4] and fiber optic sensors [5]. The existing research focused on a combination of surface metrology and image processing is very limited. In parallel to the current corrosion sensing technology, there have been a number of corrosion modeling studies attempting to numerically capture the processes of pitting corrosion initiation, pitting evolvement, pitting to cracking transition, and crack growth to fracture at the molecular level. Currently there is not a widely accepted quantitative model to take into consideration of the effect of stress factors (e.g. salinity, temperature, pressure), although the effects of the above-mentioned stress factors have been widely discussed.

Degradation detection and prediction require an appropriate estimation technique, in addition to data and a model. Estimation methods have been developed over the past years such as Kalman filtering, regression tools, etc. to address the prognosis problem. We take advantage of a novel estimation method called particle filtering that has been shown to outperform other known methods while dealing with difficult nonlinear and/or non-Gaussian problems [6]. The underlying principle of the methodology is the approximation of relevant distributions with particles (samples from the space of the unknowns) and their associated weights [6, 7]. Figure 7.6 depicts the health management scheme. Particle Filtering is an emerging and powerful methodology for sequential signal processing based on the concepts of Bayesian theory and Sequential Importance Sampling (SIS). Particle Filtering is very suitable for nonlinear systems or in the presence of non-Gaussian process/observation noise. For this approach, both diagnosis and prognosis rely upon estimating the current value of a fault/degradation dimension, as well as other important parameters, and use a set of observations (or measurements) for this purpose. This methodology involves two steps that consider the best elements of two worlds: a prediction step, based on the process model (model-based approaches) and an update step, which incorporates the new measurement into the a priori state estimate (data driven techniques).

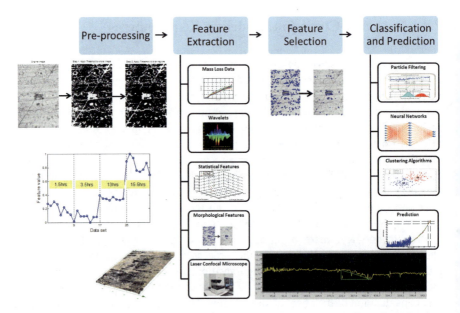

Fig. 7.6 The overall data processing, diagnostic and prognostic architecture

7.5 The Corrosion Diagnostic and Prognostic Algorithms

Figure 7.7 depicts the overall architecture for aircraft structure corrosion detection, prediction and condition based maintenance practices to intervene expeditiously and mitigate the impact of corrosion fatigue on the structural integrity of the asset.

Degradation detection and prediction require an appropriate estimation technique, in addition to data and a model, as pointed out above. Estimation methods have been developed over the past years such as Kalman filtering, regression tools, etc. to address the prognosis problem. We take advantage in this study of a novel estimation method called particle filtering that has been shown to outperform other known methods while dealing with difficult nonlinear and/or non-Gaussian problems (3). The underlying principle of the methodology is the approximation of relevant distributions with particles (samples from the space of the unknowns) and their associated weights. This is of particular benefit in diagnosis and prognosis of complex systems, because of the nonlinear behavior when operating under fault or degradation conditions.

A fault or parameter degradation diagnosis procedure involves the tasks of degradation detection, isolation and identification (assessment of the severity of the degradation). At any given instant of time, this detection framework provides a probabilistic estimate of the fault or degradation mode. Once this information is available, it is processed to generate proper fault alarms and to inform about the statistical confidence of the detection routine. Furthermore, estimates for the system

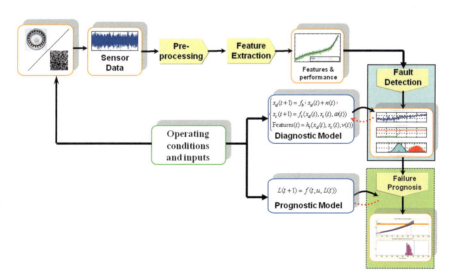

Fig. 7.7 Schematic representation of the major modules of the corrosion detection, prediction and mitigation strategies

continuous-valued states (computed at the moment of degradation detection) may be used as initial conditions in prognostic routines. Customer specifications are translated into acceptable margins for the type I and/or II errors, i.e. the false alarm and confidence or accuracy of detection.

We introduce a combined model-based and data-driven methodology that takes advantage of experimental data, corrosion progression models and an estimation method called particle filtering. It is accompanied by appropriate performance metrics while meeting customer specified requirements for detection confidence and false alarm rates, and prediction accuracy/precision.

Figure 7.8 depicts the proposed degradation diagnosis and prognosis architecture for a single mode (corrosion or no corrosion). In this architecture, real-time measurements and operating conditions are provided in real time. Data are pre-processed before computing the features that will assist to efficiently monitor the behavior of the targeted panel/coupon. With the features and a model describing the degrading state of the system, a fault detection algorithm based on particle filtering can be applied. Statistical analysis is implemented to arrive at the probability of a certain fault. When the fault is detected with a given confidence level, a failure prognostic algorithm is activated to predict the Remaining Useful Life (RUL) of the component. In our case we will be seeking to identify as accurately as possible when the corrosion level reaches specified thresholds where action might be required. This architecture provides not only a convenient compromise between data-driven and model-based techniques, but also the means to discuss its performance in terms of statistical performance indices. Moreover, a particle

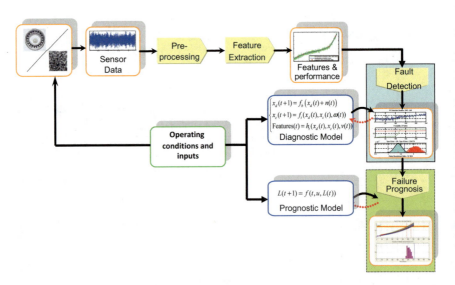

Fig. 7.8 Proposed architecture for a corrosion detector/data analysis/detection/prognosis

filtering-based algorithm enables us to efficiently deal with nonlinear systems and no-Gaussian noise-a typical situation encountered in corrosion studies.

In this architecture, the most important components supporting the implementation of the algorithm are feature extraction and diagnosis/prognosis models. Features are the foundation for "good" fault/corrosion initiation detection algorithms. Our focus is on the diagnosis and prognosis models. It is desirable to extend the architecture shown in Fig. 7.8, so that it can be used to detect and predict more than one corrosion degradation mode. There is evidence to support the contention that corrosion undergoes a series of modes before final cracking and a catastrophic failure of critical parts. For these different fault modes, as *corrosion*, *pitting* and *cracking* of aircraft aluminum alloy panels/structures, the former one usually leads to the next one. The term *corrosion* is used in this document to designate the initial corrosion stage resulting from slight pitting at the grain boundaries. The attempt is to differentiate this first observable stage from pitting where pitting corrosion is measurable at the mesoscale level.

When N degradation modes/stages are considered, due to the different degradation mechanisms, we also need N models for diagnosis and prognosis purposes. The proposed scheme is described as follows:

1. Build three models for degradation progression of corrosion, pitting and cracking. At this stage, from theoretical analysis and the literature, model structures are built and parameters are initialized for further improvement, as detailed in the chapter on corrosion modeling.
2. When new measurements (i.e. features for corrosion, pitting and cracking) become available, the parameter sets for these different models are tuned on-line.
3. Three on-line diagnosis algorithms to detect corrosion, pitting and cracking are implemented in parallel. Since detection routine carries only a very light computational burden, this will not cause real-time problems.
4. When a degradation mode is detected, the prognosis routine of this mode is activated to predict its growth as a function of time. Under the current situation, this detected mode is denoted as the "dominant" one. The prognostic routine predicts the remaining useful life of the component/part or the time to reach a predefined threshold level.
5. During the prognosis of this degradation mode, when the next t mode is detected, the dominant fault mode changes to the newly detected degradation mode. In this case, the prognosis of a newly detected mode is activated.

This way, the component/part of interest has gone through a complete diagnosis and prognosis cycle. The benefit is a more accurate and precise prediction of the component's life. In addition, the progression of corrosion stages and remaining useful life are given in an understandable way to help maintenance personnel in making correct maintenance decisions.

The fault progression is often nonlinear and, consequently, the model should be nonlinear. From a **nonlinear Bayesian state estimation** standpoint, diagnosis and prognosis are accomplished by the use of a **Particle Filter-based** module. An essential element of this module is a nonlinear state model describing the propagation of the degradation.

7.6 Corrosion Modeling Framework: Symbolic Regression

We exploit a novel modeling framework to represent stress and/or corrosion evolution in structural health management. This modeling approach allows for inclusion of impact or stress factors while accounting for uncertainty. A generic approach that can be employed to model a large variety of processes where data are available and first principle models are not feasible. The same framework is used for representing the fault evolution of critical components/systems.

Symbolic regression is a type of regression analysis that searches the space of mathematical expressions to find the model that best fits a given dataset, both in terms of accuracy and simplicity. No particular model is provided as a starting point to the algorithm. Instead, initial expressions are formed by randomly combining mathematical building blocks such as mathematical operators, analytic functions, constants, and state variables. (Usually, a subset of these primitives will be specified by the person operating it, but that's not a requirement of the technique.) New equations are then formed by recombining previous equations, using genetic programming.

By not requiring a specific model to be specified, symbolic regression isn't affected by human bias, or unknown gaps in domain knowledge. It attempts to uncover the intrinsic relationships of the dataset, by letting the patterns in the data itself reveal the appropriate models, rather than imposing a model structure that is deemed mathematically tractable from a human perspective. The fitness function that drives the evolution of the models takes into account not only error metrics (to ensure the models accurately predict the data), but also special complexity measures, thus ensuring that the resulting models reveal the data's underlying structure in a way that's understandable from a human perspective. This facilitates reasoning and favors the odds of getting insights about the data-generating system (From Wikipedia).

- Quick Recall: regression is aiming to minimize the distance between the estimated value and measured value:

$$arg \min_{f}(y - f(x))^2$$

Trying to find the best g(t)

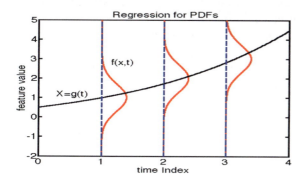

7.7 Objective Function

For each t_i, the distance (error) can be represented as follows:

$$E_i = \int_{x=-\infty}^{+\infty} (g(t_i) - x)^2 f(x, t_i)dx, \quad i = 1, 2, \ldots, N$$

Therefore, the objective function is the sum of E_i:

$$\arg\min_{g} E = \sum_{i} E_i = \sum_{i=1}^{N} \int_{x=-\infty}^{+\infty} (g(t_i) - x)^2 f(x, t_i)dx$$

7.8 Discrete Form

- We can generate samples using Monte Carlo Methods and change the distributions to samples.
- First, we assume for each time instant, t_i, the numbers of sample are the same.
- We want to generate the corresponding discrete form, which can fit our coding easier.

$$\arg\min_{g} E = \sum_{i} E_i = \sum_{i=1}^{N} \sum_{k=1}^{M} (X_{t_i,k} - g(t_i))^2$$

- Here the M is the number of samples for each distribution.
- When the number of samples for each distribution is not the same, we can use the histogram to approximate the distributions.
- The objective function is changed to the following form:

$$\arg\min_{g} E = \sum_{i} E_i = \sum_{i=1}^{N} \frac{\sum_{k=1}^{M_i} (X_{t_i,k} - g(t_i))^2}{M_i} = \sum_{i=1}^{N} \sum_{k=1}^{M_i} (\frac{X_{t_i,k} - g(t_i)}{\sqrt{M_i}})^2$$

Here the M_i is the number of samples for the distribution at time t_i.

7.9 Useful Tools

- Eureka® is a useful tool to finish the Symbolic Regression

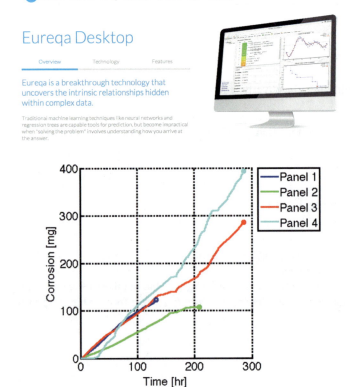

7 Corrosion Diagnostic and Prognostic Technologies

Sample Generating code

```
clc
clear all
load('C_data.mat')

N=15; % sample for each fitted
Eureqadata=[];

for i=1:360:length(time)
    if time(i)<133
        pd=fitdist(Corrosion(i,:)','normal');
    elseif time(i)<209
        pd=fitdist(Corrosion(i,2:4)','normal');
    else
        pd=fitdist(Corrosion(i,3:4)','normal');
    end
    Eureqadata=[Eureqadata;time(i)*ones(N,1),normrnd(pd.mu,pd.sigma,[1 N])'];
end
figure
hold on
for i=1:length(time)/360*N
    scatter(Eureqadata(i,1),Eureqadata(i,2));
end

save Eureqadata
```

Generated Sample Plot:

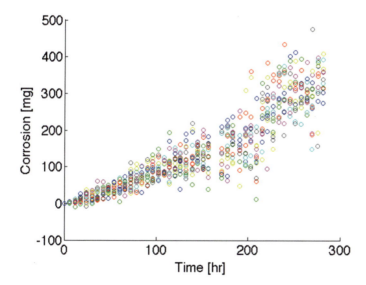

Fitting Result

- Find best solution

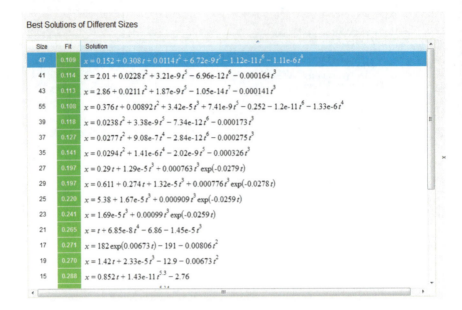

- We can choose Fit or Solution to minimize the error, or make a tradeoff between size and error.
- We do the performance analysis with the following function:

$$f1(t) = 0.152 + 0.308 * t + 0.0114 * t^2 + 6.72e-9 * t^5 - 1.12e-11 * t^6 - 1.11e-6 * t^4$$
$$f2(t) = 2.01 + 0.228 * t^2 + 3.21e-9 * t^5 - 6.96e-12 * t^6 - 0.000164 * t^3$$
$$f3(t) = 2.86 + 0.0211 * t^2 + 1.87e-9 * t^5 - 1.05e-14 * t^7 - 0.000141 * t^3$$
$$f4(t) = 182 * e^{(0.00673 * t)} - 191 - 0.00806 * t^2$$

- We test these curves with our testing data:

Fitted curve	Error
F1(t)	1.0881
F2(t)	1.0710
F3(t)	1.1029
F4(t)	1.1482

7 Corrosion Diagnostic and Prognostic Technologies

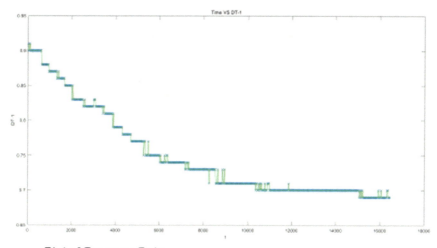

- Plot of Pressure Data.

Fig. 7.9 Pressure data

What do we need?

- Area percentage feature or some other global feature act as the ground truth.
- After we have the relationship between corrosion and time, we will eliminate t to get corrosion versus feature (Figs. 7.9 and 7.10).

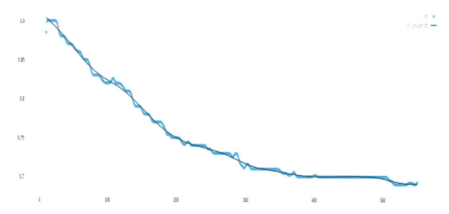

Fig. 7.10 Model prediction and smoothing

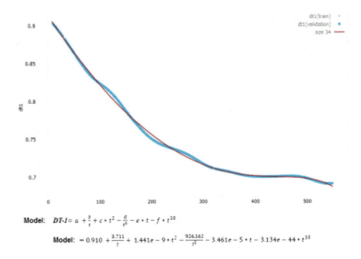

Model: $DT\text{-}1 = a + \frac{b}{t} + c*t^2 - \frac{d}{t^2} - e*t - f*t^{10}$

Model: $= 0.910 + \frac{3.711}{t} + 1.441e-9*t^2 - \frac{926.142}{t^2} - 3.461e-5*t - 3.134e-44*t^{10}$

7.10 Symbolic Regression Result

$$\text{Model}: DT - 1 = a + \frac{b}{t} + c*t^2 - \frac{d}{t^2} - e*t - f*t^{10}$$

$$\text{Model}: = 0.910 + \frac{3.711}{t} + 1.441e - 9*t^2 - \frac{926.162}{t^2} - 3.461e - 5*t - 3.1342e - 44*t^{10}$$

An Integrating End-to-End Architecture for Corrosion Diagnosis and Prognosis

We introduce a rigorous and verifiable framework for diagnosis and prognosis, developed, tested and applied to various laboratory, military and commercial systems that builds upon a **systems engineering process** as the driving philosophy for health management [6]. The online modules (see Fig. 7.6) perform raw data pre-processing, feature extraction, fault diagnosis and failure prognosis that exploit available ground truth fault data, noise models, experimental data, system models and other tools offline to tune and adapt online parameters and estimate suitable mappings.

Physics of Failure Mechanisms

The foundation for the development and application of PHM technologies is a thorough understanding of the physics of failure mechanisms as critical systems are subjected to stress conditions. From the physical components/systems themselves to a good understanding of how such systems fail and under what conditions leads to optimum Condition Indicator (CI) extraction and selection and, eventually, to accurate diagnostics and prognostics.

7 Corrosion Diagnostic and Prognostic Technologies

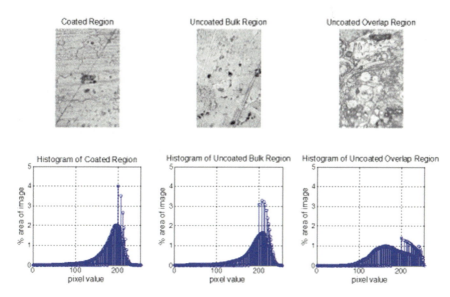

Fig. 7.11 Image processing and histogram generation

Failure Modes and Effects Criticality Analysis (FMECA)

The starting point for "good" diagnostics/prognostics is a thorough FMECA. It describes the failure modes, sensor suite, condition indicators, possible diagnostics and prognostic algorithms. It forms the first essential step in the systems engineering process for health management of critical aircraft components/systems (Fig. 7.11).

Sensors and Sensing Strategies

Sensors and sensing strategies constitute the essential requirements for fault diagnosis and failure prognosis of failing components/systems. The type, location and characteristic properties of PHM sensors, i.e. sensors that are specifically designed to monitor corrosion degradation signatures, present major challenges to the system designer. Figure shows typical sensor results with devices embedded to monitor corrosion processes. We introduce an approach to determine the type, number and location of sensing modalities that maximize the fault signal to noise ratio (Fig. 7.12).

Data Pre-processing

Raw sensor data (primarily images of corroding surfaces, temperature, humidity, etc.) must be pre-processed in order to reduce the data dimensionality and improve the (degradation) Signal to Noise Ratio (SNR). Typical pre-processing routines include data compression and filtering, Time Synchronous Averaging (TSA) of time series data, FFTs, wavelet decomposition techniques, among others.

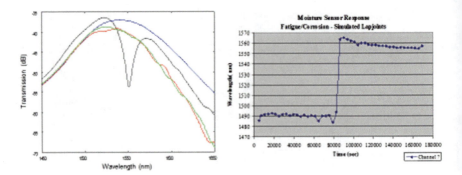

Fig. 7.12 Temperature and moisture sensing using embedded sensors

Data Mining

Data mining is "the non-trivial extraction of implicit, previously unknown, and potentially useful information from data". Data mining tools typically perform an exploratory data analysis on large data sets; they focus on understandability rather than accuracy or predictability, and they are pattern focused rather model-focused. We propose a data mining process that involves the following steps: Learning the application domain, data cleaning and preprocessing, data reduction (PCA and other tools will be called upon to reduce the data dimensionality and identify possible data redundancies). We propose a systems engineering process to analyze the targeted data set, to reduce the data dimensionality and identify meaningful data clusters that will facilitate the task of meta-data design, i.e. to determine appropriate work tasks, evaluate these tasks and, ultimately, decide upon potential benefits. Data mining tools include Bayesian-based approaches, Inductive Decision Trees, tools from the computational intelligence domain, such as fuzzy logic or neuro-fuzzy constructs, neural networks, among others, Principal Component Analysis (PCA).

Metadata Design The metadata design effort aims to consolidate the data mining results and assist in determining appropriate design actions (as dictated from the modeling and optimization functions), evaluate these actions and, ultimately, decide upon potential benefits. Towards this effort, we call upon dynamic reasoning tools from the Machine Learning domain. These techniques are based on well-known numerical algorithms that have found numerous applications in signal processing and data mining, among others [8, 9]. We will introduce and demonstrate an adaptive form of these algorithms that updates the solution as new data become available. Missing value estimation techniques will be employed to handle the cases when not all of the data are available. Imaging data constitute the basic sensing output exploited for corrosion detection and prediction. Figure 7.13 shows the imaging steps from pre-processing to feature extraction and classification. Statistical and morphological features are extracted using the wavelet decomposition technique.

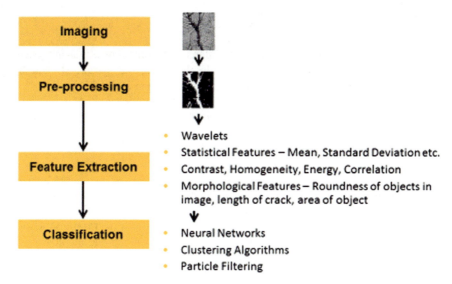

Fig. 7.13 The data imaging, feature extraction and classification framework

7.11 Health Indexes

We define and use in the development of the health management architecture Health Indexes (HIs), concepts that are useful to assess the current and future health status of corroding surfaces. The HI definition begins with a generalized equation for wear/fault growth/degradation of the form:

$$w = Ae^{B(t)}$$

Which retains the macro-level degradation characteristics. Next, we assume an upper wear threshold $th(w)$ that denotes an operational limit beyond which the component/subsystem cannot be used. The generalized wear equation can be rewritten as a time varying health index, $h(t)$, by subtracting wear from the upper wear threshold and normalizing it with respect to the upper wear threshold as

$$h(t) = 1 - \frac{Ae^{B(t)}}{th(w)}$$

This equation can be rewritten as

$$h(t) = 1 - e^{ate^b}$$

by recasting the parameters. Generally, the system will be observed with some initial degradation, d, which is modeled as an additive term resulting in

$$h(t) = 1 - d - e^{ate^b}$$

The health index can be used to model different phenomena within a subsystem—the identified components in our case. In this case, the system HI can be described as the minimum of all operative margins written as

$$g(e(t), f(t), l(t), \ldots) = \min(m_1, m_2, m_3, \ldots)$$

The margins m_1, m_2, etc. are functions of the specific fault modes considered.

Condition Indicator Extraction and Selection

Condition Indicator (CI) selection and extraction constitute the cornerstone for accurate and reliable corrosion degradation diagnosis. The objective is to transform high dimensional raw data into tractable low dimensional form (information) without loss of useful information. The objective is to develop and employ novel feature extraction and selection methodologies that establish the foundational elements for accurate fault diagnosis while processing large volumes of raw data to arrive at useful information. For this purpose, we explore the utility of appropriate metrics in on-line and real-time feature extraction algorithms. Over the past several years researchers at Georgia Tech have investigated multiple feature extraction and selection algorithms and defined suitable performance metrics, such as z-score and correlation coefficient [10, 11]. Building on this experience, we develop a sound theoretical foundation based on the data mining notions suggested in Chapter. Of special interest are novel feature extraction tools based on Deep Learning (Deep Learned Features) and smart control algorithms for selection and classification purposes.

Degradation/Fault diagnosis depends mainly on extracting a set of features from sensor data that can distinguish between corrosion classes/stages of interest, detect and isolate a particular degradation mode at its early initiation stages. In selecting an "optimum" feature set, we are seeking to address such questions as: Where is the information? How do degradation (failure) mechanisms relate to the fundamental "physics" of complex systems? Fault/Degradation modes may induce changes in the energy, entropy, power spectrum, signal magnitude, etc. [10]. We propose to pursue a data mining formalism in order to determine the "best" features that are descriptive of the faulty behavior of critical components/subsystems. Kolmogorov Complexity is viewed by the data mining community as the foundation for algorithms that aim to process large volumes of data (data, images, observations), i.e. perform such functions as data compression, clustering, classification, anomaly detection, forecasting, etc. Feature extraction and selection is indeed at the heart of the data mining problem. Borrowing concepts from this area will help to formalize an approach that has relied thus far on heuristics, intuition and the best judgment of experts.

For purposes of computing features and selecting the optimum feature vector for diagnosis and prognosis we define the Compression-Based Dissimilarity Measure as:

$$CDM = \frac{C(xy)}{C(x)+C(y)}$$

where, in the feature extraction algorithm $C(x)$ is the size of the compressed x, and $C(x|y)$ is the compression achieved by first training the compression on y (ground truth data), and then compressing x. Typical compressors are based on textual substitution methods, among others. The CDM dissimilarity is close to one when x and y are not related, and smaller than one if x and y are related. The smaller the $CDM(x, y)$, the more closely related x and y are.

The objective is to develop and employ novel feature extraction and selection methodologies that establish the foundational elements for accurate fault diagnosis while processing large volumes of raw data to arrive at useful information. For this purpose, we explore the utility of appropriate metrics in on-line and real-time feature extraction algorithms. Over the past several years researchers at Georgia Tech have investigated multiple feature extraction and selection algorithms and defined suitable performance metrics, such as z-score and correlation coefficient [10, 11]. Building on this experience, we develop a sound theoretical foundation based on the data mining notions suggested in Chapter. Of special interest are novel feature extraction tools based on Deep Learning (Deep Learned Features) and smart control algorithms for selection and classification purposes.

7.12 Development of Fault Diagnosis and Failure Prognosis Algorithms

The objective is to develop a sequence of methods and tools that establish the mathematical rigor, effectiveness and desired performance of the new and enhanced diagnostic and prognostic algorithms. The fault/degradation progression is often nonlinear and, consequently, the model should be nonlinear. From a nonlinear Bayesian state estimation standpoint, diagnosis and prognosis may be accomplished by the use of a Particle Filter-based module. An essential element of this module is a nonlinear state model describing the propagation of the degradation.

It is structured in terms of a number of subtasks, each with a dedicated function and goal, used as building blocks (pieces) in the overall diagnostic/prognostic/reconfigurable control mosaic. We propose to address the development, design, testing and implementation of new model-based and data-driven diagnostic and prognostic algorithms. Bayesian estimation methods, and particularly particle filtering, are the underpinnings of robust and accurate diagnosis and prognosis. The task here is to suggest enhancements to algorithms that have shown already to outperform existing or state-of-the-art routines and introduce new model-based and data-driven methods that improve further the effectiveness of these tools. Figure 7.8 shows the particle filtering formulation for degradation detection and prediction. This framework has been implemented in real time on critical aerospace systems exhibiting verifiable response characteristics. Its generic aspects are applicable to on-line corrosion detection and prediction.

The proposed corrosion diagnosis and prognosis framework builds upon mathematically rigorous concepts from estimation theory—an emerging and powerful methodology in Bayesian theory called Particle Filtering that is particularly useful in dealing with difficult non-linear and/or non-Gaussian problems. Particle filtering facilitates the estimation of the state (fault) model over consecutive time instants as measurements become available. The particle filtering routines for diagnosis and prognosis are implemented and executed in near real-time and constitute an integrated framework where the results of diagnosis serve as the initial conditions for prognosis in a transparent and efficient manner.

Particle filtering is an emerging and powerful methodology for sequential signal processing based on the concepts of Bayesian theory and Sequential Importance Sampling. The underlying principle of the methodology is the approximation of relevant distributions with particles (samples from the space of the unknowns) and their associated weights. Compared to classical Monte Carlo methods, sequential importance sampling enables particle filtering to reduce the number of samples required to approximate the distributions with necessary precision, and makes it a faster and more computationally efficient approach than Monte Carlo simulation. It is very suitable for nonlinear systems and/or in the presence of non-Gaussian process/observation noise. One particular advantage of the proposed particle filtering approach is the ability to characterize the evolution in time of the nonlinear fault (state) model through modification of the probability masses associated with each particle, as new CI or feature information is received.

The Georgia Tech research team has pioneered the introduction of particle filtering techniques into fault diagnosis and failure prognosis [12, 6]. The success of this novel approach has been demonstrated in a number of diverse application domains from rotorcraft critical components to electrical systems, environmental control systems and high power amplifiers [13, 14]. We will introduce a new approach to particle filtering as a model-based technique in combination with a data-driven mapping for improved prediction accuracy and precision. This is of particular benefit in diagnosis and prognosis of complex dynamic systems, such as gearboxes, drive systems/components, etc. Moreover, particle filtering allows for information from multiple measurement sources to be fused in a principled manner.

From a particle filtering perspective, real-time fault/degradation diagnosis and prognosis is aiming to estimate recursively the current/long-term corrosion states by taking into account available measurements, while both diagnosis and prognosis contain two key steps: prediction and update. The prediction is intended to estimate the prior pdf of the states by using the nonlinear system (fault) model, while the update step involves modifying the prior density to gain the posterior density through the use of an appropriate measurement model. Long-term prediction of the

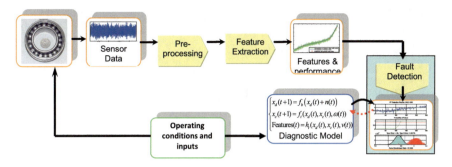

Fig. 7.14 Proposed architecture for the fault (anomaly) detector

fault evolution will be based on an estimate of the current state and a model describing the fault progression, more specifically the fault-growth model. Uncertainty associated with long-term predictions is managed by using the current state pdf estimate, process noise model, and a record of corrections made to previously computed predictions.

In prognosis, defining an initial estimate of the fault pdf is a crucial step since the long-term prediction accuracy and precision are strongly dependent on this initial estimate and tend to deteriorate rapidly with inaccurate initial conditions. In the proposed scheme, the initial pdf is defined automatically by the final diagnostic outcome, assuring a satisfactory degree of accuracy/precision, avoiding an ad hoc choice for this estimate and preserving a transparent continuity between fault diagnosis and failure prognosis resulting in improved algorithm effectiveness. We will introduce appropriate performance metrics to assess the effectiveness of the diagnostic/prognostic routines and to assist in their optimum design.

7.13 Corrosion Degradation Detection—The Particle Filtering Approach to Degradation Detection

Fault or degradation Diagnosis—Detection, isolation and identification of an impending or incipient failure condition—the affected component (subsystem, system) is still operational even though at a degraded mode.

Figure 7.14 depicts the major modules of the proposed architecture for a fault detector.

Figure 7.15a shows a Simulink implementation of the modules depicted in Fig. 7.14. Features or Condition Indicators are extracted in the frequency domain and a neural network based classifier, as shown in Fig. 7.15b, maps the selected/extracted features into a fault dimension.

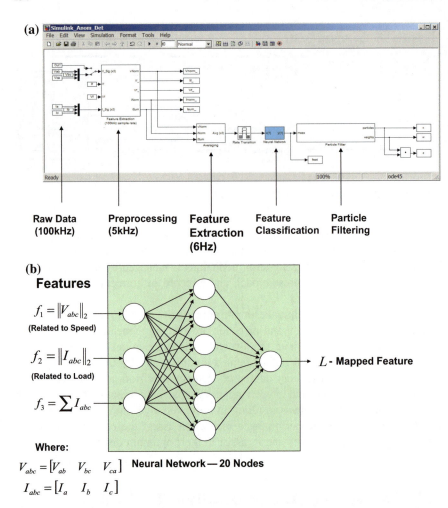

Fig. 7.15 a Simulink implementation of anomaly detection routine. **b** Feature classification

The NN is trained and a typical residual distribution is shown in Fig. 7.16. The particle filter Simulink implementation block is shown in Fig. 7.17. Results under no-fault and fault conditions for a typical motor winding degradation test are shown in the watershed schematics of Fig. 7.18. These results are encouraging and suggest that potential risks associated with the proposed algorithms are mitigated successfully.

7 Corrosion Diagnostic and Prognostic Technologies

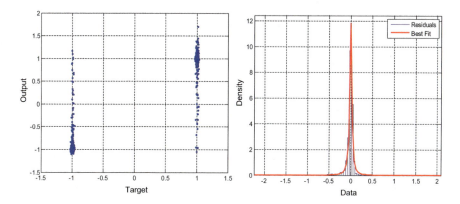

Fig. 7.16 Residual distribution

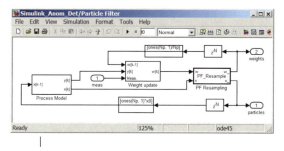

Process Model

$$\begin{cases} L[k+1] = sat(L[k] + \gamma L[k](L[k] - L[k-1])) + w[k] \\ y[k] = L[k] + v[k] \end{cases} \text{ where } sat(x) = \begin{cases} x & |x| < 1 \\ \text{sgn}(x) & |x| \geq 1 \end{cases}$$

Fig. 7.17 Particle filtering Simulink implementation

7.14 Diagnosis of Corrosion Degradation

The corrosion detection and identification procedure may be interpreted as the fusion and utilization of the information present in a feature vector (observations) with the objective of determining the operational condition (state) of a system and the causes for deviations from particularly desired behavioral patterns [5].

A diagnosis procedure for early corrosion detection involves the tasks of detection and identification (assessment of the severity of the corrosion). In this sense, the proposed particle-filter-based diagnostic framework aims to accomplish these tasks, under general assumptions of non-Gaussian noise structures in the measurements and corrosion process, nonlinearities in process dynamic models,

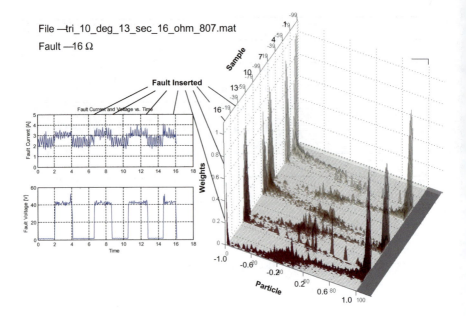

Fig. 7.18 Test results under fault condition

using a reduced particle population to represent the state probability density function (PDF) [15]. The particle filter-based module builds on the nonlinear dynamic state model (corrosion growth),

$$\begin{cases} x_d(t+1) = f_b(x_d(t), n(t)) \\ x_c(t+1) = f_t(x_d(t), x_c(t), w(t)) \\ f_p(t) = h_t(x_d(t), x_c(t), v(t)) \end{cases} \quad (7.1)$$

where f_b, f_t and h_t are non-linear mappings, $x_d(t)$ is a collection of Boolean states associated with the presence of a particular operating condition in the system (normal operation, fault type #1, #2, etc.), $x_c(t)$ is a set of continuous-valued states that describe the evolution of the system given those operating conditions, $f_p(t)$ is a feature measurement, $w(t)$ and $v(t)$ are non-Gaussian distributions that characterize the process and feature noise signals respectively. At any given instant in time, this framework provides an estimate of the probability masses associated with each fault mode, as well as a PDF estimate for meaningful physical variables in the system. The FDI module generates proper fault alarms and as well as the statistical confidence of the detection routine. Performance metrics are translated into acceptable margins for the type I (false positives) errors and type II errors (false negatives) in the detection routine. The algorithm itself will indicate when the type II error has decreased to the desired level. Figure 7.19 shows an application of diagnostic tools to pipeline corrosion monitoring.

7 Corrosion Diagnostic and Prognostic Technologies

- **Issues**
  - Detect defects
  - Polar environment
  - Overlapping defects
  - Continuous data stream

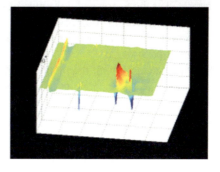

Fig. 7.19 Pipeline diagnostics

The particle-filter-based diagnosis framework aims to accomplish the tasks of fault detection and identification using a reduced particle population to represent the state probability density function (pdf). This framework provides an estimate of the probability masses associated with each fault mode, as well as a pdf estimate for meaningful physical variables in the system. Once this information is available within the diagnostic module, it is conveniently processed to generate proper fault alarms and to inform about the statistical confidence of the detection routine. Customer specifications are translated into acceptable margins for the type I and II errors in the detection routine. The algorithm itself will indicate when the type II error (false negatives) has decreased to the desired level.

7.15 Corrosion Diagnosis—Implementation Issues

Particle filtering is used in diagnosis routines to detect with a specified level of confidence the initiation of corrosion. The corrosion detection method using particle filtering is outlined below:

Step 1: Initialize particle filter

- Generate probability density function (PDF) from baseline data.
- From baseline data generate initial values of particles.

Step 2: Run particle filter

- Update particles using model.
- Update weights for each particle.

- Get current PDF of the fault dimension from particles and weights.
- Determine detection confidence based on current PDF, baseline PDF and given false alarm rate. Figure 7.15 demonstrates how to compute the detection confidence. The false alarm rate determines which point on the x-axis to start integrating the current PDF to compute the detection confidence.

The Fault/Degradation Detection and Identification (FDI) procedure may be interpreted as the fusion and utilization of the information present in a feature vector (observations) with the objective of determining the operational condition (state) of a system and the causes for deviations from particularly desired or normal behavioral patterns.

For an actual system, the model can be written in the following form:

$$\begin{bmatrix} x_{d,1}(t+1) \\ x_{d,2}(t+1) \end{bmatrix} = f_b\left(\begin{bmatrix} x_{d,1}(t) \\ x_{d,2}(t) \end{bmatrix} + n(t)\right)$$
$$x_c(t+1) = [(1+\beta)x_c(t)]x_{d,2}(t) + w(t)$$
$$y(t) = x_c(t) + v(t)$$
$$f_b(x) = \begin{cases} [1\ 0]^T, & if\ ||x - [1\ 0]^T|| \leq ||x - [0\ 1]^T|| \\ [0\ 1]^T, & else \end{cases}$$
$$[x_{d,1}(0)x_{d,2}(0)x_c(0)] = [1\ 0\ 0]$$

In this model, f_b is a non-linear mapping, $x_{d,1}$ and $x_{d,2}$ are Boolean states that indicate normal and faulty conditions, respectively, $y(t)$ is the noise-contaminated fault dimension, and β is a time-varying model parameter that describes the propagation of the fault dimension under current operating conditions. Figure 7.20 is a schematic representation of the detection scheme.

Figure 7.21 details the definition of the false alarm rate and detection confidence or accuracy. Typically the false alarm rate is specified a priory, i.e. 5% or 2%, and the confidence is estimated from the overlap region between the baseline pdf and the one computed during the current analysis procedure.

The degradation detection and identification procedure may be interpreted as the fusion and utilization of the information present in a feature vector (observations) with the objective of determining the operational condition (state) of a system and

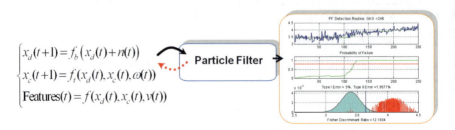

Fig. 7.20 Particle filtering—a fault/degradation detection method

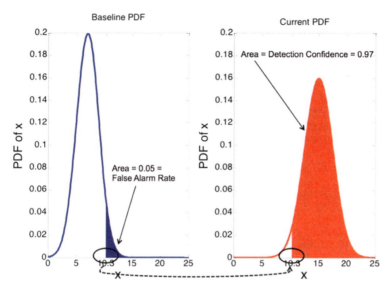

Fig. 7.21 False alarm rate and computing detection confidence

the causes for deviations from particularly desired behavioral patterns. A model for diagnosis is given Eq. (7.1). Model-based techniques for fault/degradation diagnosis are capable of addressing unanticipated degradation conditions if an accurate corrosion process is available. Data-driven routines are employed to address anticipate faults, if a statistically sufficient data base is available. Hybrid configurations using both data-driven and model-based tools/methods have been reported in the literature that aim to improve the diagnosis and prognosis results taking advantage of both techniques. Figure 7.22 illustrates the two approaches.

The particle-filter-based diagnosis framework aims to accomplish the tasks of degradation detection and identification using a reduced particle population to represent the state probability density function (pdf). This framework provides an estimate of the probability masses associated with each fault mode, as well as a pdf estimate for meaningful physical variables in the system. Once this information is available within the diagnostic module, it is conveniently processed to generate proper fault alarms and to inform about the statistical confidence of the detection routine. Customer specifications are translated into acceptable margins for the type I and II errors in the detection routine. The algorithm itself will indicate when the type II error (false negatives) has decreased to the desired level. Figure 7.23 shows the anomaly detection results based on an RMS feature. The first plot depicts the progression of the feature as a function of time while the second is the probability of failure; the last one shows the baseline and fault pdfs at 5% false alarm rate. The Type II error is 1.1117% at that specific instant of time. Another performance metric is the Fisher Discriminant Ratio shown at the bottom of Fig. 7.23.

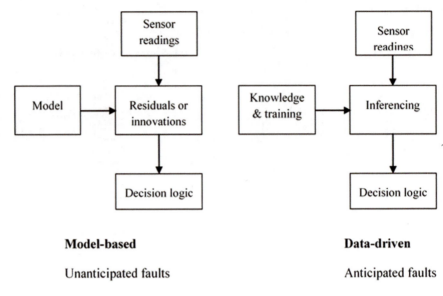

Fig. 7.22 Model based and data driven diagnostic method

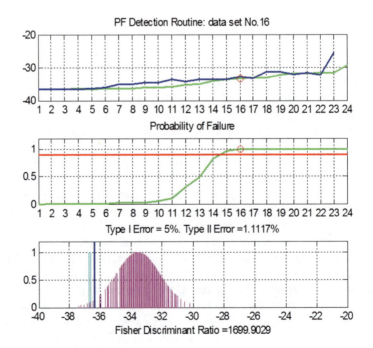

Fig. 7.23 Typical diagnostic results

Corrosion Detection and Quantification Using Image Processing for Aging Aircraft

Localized corrosion is recognized as one of the degradation mechanisms that affect the structural integrity of aging aircraft structures. Several nondestructive inspection (NDI) systems (eddy current, ultrasound and others) have been used to obtain the images of damaged regions. There is a growing demand for improving existing NDI techniques to achieve maximum confidence and reliable results with minimum damage components. There is always a constant outlook for methods that identify the damaged regions on the image and also that gives a quick estimate of the extent of the damage.

The images obtained through conventional NDI methods are not directly suitable for identification and quantification of damaged regions. Such images therefore need to be enhanced and segmented appropriately for further image analysis. In this paper, segmentation has been achieved using wavelet decomposition. The wavelet coefficients also can be used to quantify the extent of corrosion. Neural networks were applied in the process of segmentation and quantification of damaged regions. Segmentation results show a good correspondence between the extracted regions and the actual damage on sample panels. A good accuracy was obtained in distinguishing the corroded segments from the non-corroded ones. An accuracy of about 75% was achieved in the quantification of the corroded panels.

Corrosion Detection and Quantification using Image Processing for Aging Aircraft

Localized corrosion is recognized as one of the degradation mechanisms that affect the structural integrity of aging aircraft structures. Several nondestructive inspection (NDI) systems (eddy current, ultrasound and others) have been used to obtain the images of damaged regions. There is a growing demand for improving existing NDI techniques to achieve maximum confidence and reliable results with minimum damage components. There is always a constant outlook for methods that identify the damaged regions on the image and also that gives a quick estimate of the extent of the damage.

The images obtained through conventional NDI methods are not directly suitable for identification and quantification of damaged regions. Such images therefore need to be enhanced and segmented appropriately for further image analysis. In this paper, segmentation has been achieved using wavelet decomposition. The wavelet coefficients also can be used to quantify the extent of corrosion. Neural networks were applied in the process of segmentation and quantification of damaged regions. Segmentation results show a good correspondence between the extracted regions and the actual damage on sample panels. A good accuracy was obtained in distinguishing the corroded segments from the non-corroded ones. An accuracy of about 75% was achieved in the quantification of the corroded panels.

Intelligent Computational Methods for Corrosion Damage Assessment

Corrosion is one of the major damage mechanisms affecting the structural integrity of aging aircraft structures. Various nondestructive inspection (NDI) techniques are being used to obtain images of corroded regions on structures. This paper describes a computational approach using Wavelet transforms and artificial neural networks to analyze and quantify the extent of corrosion damage from the NDI images. The Wavelet parameters obtained from the images were first used to classify between corroded and un-corroded regions using a clustering algorithm. The corroded regions were further analyzed to obtain the material loss due to corrosion using artificial neural network model. Experiments were carried out to investigate the developed methods for aircraft panels with engineered corrosion obtained from the FAA Validation Center in Albuquerque. The results presented indicate that the computational methods developed for corrosion analysis seem to provide reasonable results for estimating material loss due to corrosion damage.

Use of Computational Intelligence for Corrosion Damage Assessment of Aircraft Panels

Localized corrosion is recognized as one of the degradation mechanisms that affect the structural integrity of aging aircraft structures. Several nondestructive inspection (NDI) systems (eddy current, ultrasound and others) have been used to obtain the images of damage regions. There is a growing demand for improving existing NDI techniques to achieve maximum confidence and reliable results with minimum damage components. There is always a constant outlook for methods that identify the damaged regions on the image and also that gives a quick estimate of the extent of the damage.

The images obtained through conventional NDI methods are not directly suitable for identification and quantification of damaged regions. Such images therefore need to be enhanced and segmented appropriately for further image analysis. In this paper, segmentation has been achieved using wavelet decomposition. The wavelet coefficients also can be used to quantify the extent of corrosion. Neural networks were applied in the process of segmentation and quantification of damaged regions. Segmentation results show a good correspondence between the extracted regions and the actual damage on sample panels. A good accuracy was obtained in distinguishing the corroded segments from the non-corroded ones. An accuracy of about 75% was achieved in the quantification of the corroded panels.

7.16 Beyond Diagnosis Towards Prognosis

The proposed segment-by-segment approach provides ample observations for the propagation of local anomaly in the time domain. This feature can be exploited to enable data-driven prognostic analysis beyond diagnostic anomaly detection. A fundamental challenge in prognosis stems from the "large-grain" uncertainty inherent in the prediction task. Long-term prediction of the NAS degradation evolution, to the point that may result in a failure or detrimental event requires means to represent and manage the inherent uncertainty. Data uncertainty, process (system) uncertainty, environmental uncertainty, measurement and modeling uncertainties are potential contributors to prognostic uncertainty in aircraft systems. The prognosis scheme should consider critical state variables (such as degradation indicators) as random processes in such a way that, once their probability distributions are estimated, other important attributes (such as confidence intervals) may be computed.

An important distinction must be drawn between two major categories of prognostic algorithms: Health-Based versus Usage-Based prognostics. The former refers to prognostic approaches developed and applied on-line in real time as the system (NAS) at hand is monitored and data are streaming into a processor for degradation detection and failure prognosis. In this case, an incipient failure is detected first with specified confidence and then the prognostic algorithm is initiated to predict the time evolution. The final degradation state acts as the initial condition for prognosis. In contrast, usage-based prognosis considers the past, current and assumed future usage or stress patterns of the NAS system to estimate the system's remaining useful life. Such prognostic methods do not presuppose the existence of incipient failure modes in deference to health based prognostics. The remaining useful life may be estimated at any time in the system's operating history in the absence of a degradation condition. The prediction may be continuously updated as new evidence accumulates. Life cycle management tools for critical NAS systems take advantage of usage-based prognostic routines to arrive at times needed for corrective actions. The enabling technologies include neuro-fuzzy systems, response surface methodologies, similarity based methods and regression analysis techniques, among others.

7.17 Degradation Prognosis

Prognosis is the ability to predict accurately and precisely the remaining useful life (RUL) or time for remediation of a degrading structure/part.

Prognosis is the Achilles' heel of degradation diagnosis and failure prognosis systems. Prognosis can be understood as the generation of long-term predictions describing the evolution in time of a particular signal of interest or fault/degradation indicator. Since prognosis projects the current condition of the indicator in the absence of future measurements, it necessarily entails large-grain uncertainty. This suggests a prognosis scheme based on recursive Bayesian estimation techniques, combining both the information from fault growth models and on-line data obtained from sensors monitoring key degradation parameters (observation or features). Proposed is a prognostic framework that takes advantage of a nonlinear process (fault/degradation) model, a Bayesian estimation method using particle filtering and real-time measurements.

Prognosis is achieved by performing two sequential steps, prediction and filtering. Prediction uses both the knowledge of the previous state estimate and the process model to generate the a priori state pdf estimate for the next time instant, It may be understood as the result of the procedure where long-term (multi-step) predictions—describing the evolution in time of a parameter or degradation indicator—are generated with the purpose of estimating the Remaining Useful Life (RUL) or time to take appropriate action. The same particle filtering framework and nonlinear state model suggested above will be used to estimate the RUL [16, 17]. The state estimation is achieved recursively in two steps: prediction and update. The prediction step is intended to obtain the prior probability density function of the state for the next time instant.

The detailed algorithm steps for condition prognosis may be stated as:

Step 1: The Symbolic Regression model is trained with available condition data to model the degradation propagation process.

Step 2: The degradation growth model is employed in the particle filtering formulation to draw a set of particles. According to the values of the particles and current weights, condition prediction is carried out next.

When a new measurement becomes available, the weights of the particles or samples are calculated.

Step 3: Update the process noise and model parameters.

Step 4: Repeat *Step* 2 and *Step* 3 until prognosis is complete.

Elements for Prognostics

There are some basic ingredients that are common to all prognostic approaches. These are: a model that describes both the system under investigation and damage propagation, a quantification of the damage threshold, an algorithm that handles the propagation of the damage/degradation into the future, and a mechanism to deal with uncertainty.

The system model describes the characteristics of the system under nominal conditions. Ideally, such a model should be able to factor in the effects of operational and environmental conditions as well as any other conditions that cause different system response under nominal conditions. The system model could integrate domain expertise and be implemented using rules, a physical description of the system under investigation or it could learn system behavior from examples, for instance using machine learning techniques.

A damage/degradation propagation model describes how the degradation is expected to grow in the future. It should, similar to the system model, account for operational and environmental conditions as well as any other conditions that have an impact on the damage. While one often thinks of damage as a monotonically increasing phenomenon, it is possible for the domain in which damage is evaluated to have non-monotonic attributes. The prognostic algorithm applies the damage propagation model into the future. The damage propagation algorithm needs to ensure that it properly considers the effects of environmental and operational conditions.

The damage threshold needs to be established since one needs to know what condition should terminate the end-of-life prediction. A requirement for an end-of-life threshold in prognostics is a measurable condition. This is not always the same conditions as a catastrophic event. Often, these are performance specifications that constrain the operation of the component. It is possible that the system may continue to operate beyond the limits of the end-of-life conditions (but outside of the specifications).

Prognostics is not really useful unless the uncertainties in the predictions are accounted for. Uncertainty management tools seek to improve the signal (degradation) to noise (uncertainty) ratio. They begin by determining the uncertainty sources in terms of an uncertainty tree and then exploit filtering or kernel-based methods for uncertainty management [18]. A remaining life estimate that has no quantification of the certainty bounds leaves the user with little practical information. Accounting for the various sources of uncertainty and rigorously combining them will allow decision makers to justify the action taken. Uncertainties need to be managed carefully because a haphazard stacking of the various uncertainties may lead to wide bounds that wipe out the benefits of estimating remaining life. Sources of uncertainty can come from the models (both their structure as well as their parameters), the current state estimate, the future NAS state and environmental conditions.

Prognostic Algorithm Approaches for Corrosion Prediction

In the engineering disciplines, failure prognosis has been approached via a variety of techniques ranging from Bayesian estimation and other probabilistic/statistical methods to artificial intelligence tools and methodologies based on notions from the computational intelligence arena [19]. Specific enabling technologies include multi-step adaptive Kalman filtering [20], auto-regressive moving average models, stochastic auto-regressive integrated moving average models [21], Weibull models [22], forecasting by pattern and cluster search [23], parameter estimation methods [24], and particle filter methods [16]. From the artificial intelligence domain, case-based reasoning [25], intelligent decision-based models and min-max graphs have been considered as potential candidates from prognostic algorithms. Other methodologies, such as Petri nets, neural networks, fuzzy systems and neuro-fuzzy systems [26] have found ample utility as prognostic tools as well. A comprehensive review of computational intelligence methods for prognostics is given in Schwabacher and Goebel [27]. Physics-based fatigue models [28, 29] have been extensively employed to represent the initiation and propagation of structural anomalies.

Failure Prognosis Algorithms The role of the prognostic algorithm is projecting the damage propagation model into the future [19]. The algorithm needs to ensure that it properly considers the effects of environmental and operational conditions. It is a stochastic model with variables and parameters expressed as Probability Density Functions (PDFs). We have developed and applied to various engineering systems a variety of prognostic algorithms [6]. Results from these studies were compared with our preferred method—particle filtering, developed by this research team for failure prognosis purposes. In all cases, particle filtering, as a Bayesian estimation method outperformed other approaches. We will rely initially on particle filtering for prognostic purposes.

Damage Threshold Damage thresholds define the end-of-life, or failure condition. An end-of-life threshold in prognostics must be a measurable condition, although it should be noted that this condition is not always the complete destruction of a component. In fact, to be effective there should always be some margin between this threshold and complete destruction and the system may continue to operate beyond the end-of-life threshold conditions The damage threshold is also designated as the "hazard zone"; it represents the anticipated end of life as a PDF. Various hazard zones or threshold levels may be allocated to suggest degrees of confidence in the end of life estimates. Specifying such hazard zones requires actual component failure data—missing in most cases—presenting an additional challenge to the designer.

Next, we provide a brief overview of a representative sample of the multitude of enabling technologies. Prognostic technologies typically utilize measured or inferred features, as well as data-driven and/or physics-based models, in conjunction with an estimation method, to predict the condition of the system at some future time. Inherently probabilistic or uncertain in nature, prognostics can be applied to failure modes governed by material conditions or by functional loss. Prognostic algorithms can be generic in design but are typically rather specific when used in the context of a particular application. Prognostic system developers have implemented various approaches and associated algorithmic libraries for customizing applications that range in fidelity from simple historical/usage models to approaches that utilize advanced feature analysis or physics-of-failure models.

Various approaches are needed to develop and to implement the desired prognostic approach depending on (besides resource availability) the criticality of the Least Replaceable Unit (LRU) or subsystem being monitored but also on the availability of data, models, and historical information. Figure 7.24 summarizes the range of possible prognostic approaches as a function of the applicability to various systems and their relative implementation cost. The pyramid starts at the base with generic, statistical life usage and experience-based prognostic models, migrates to data-driven techniques employing evolutionary or trending models while physics-based models employed for prognostic purposes occupy the top of the pyramid. Prognosis technologies typically use measured or inferred features, as well as data-driven and/or physics-based models, to predict the condition of system at some future time [16, 19]. Model-based prognostics, at the top of the pyramid, should guide the future development of reliable and verifiable prognostic algorithms for complex systems such as the NAS.

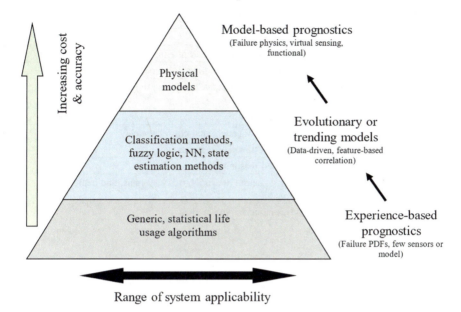

Fig. 7.24 A taxonomy of prognostic approaches

7.18 A Taxonomy of Prognostic Approaches

The pyramid in Figure describes a taxonomy of prognostic approaches starting with generic, statistical life usage algorithms at the base and moving up to classification methods borrowing from the Neural Net, fuzzy logic and state estimation techniques; physical models and model-based prognostic routines occupy the top of the pyramid.

The book "*System of Health Management: with Aerospace Applications*" [30] reviews the variety of approaches of fault prognosis in the engineering discipline. Techniques ranging from Bayesian estimation and other probabilistic/statistical methods to artificial intelligence tools have been applied. Table 7.1 summarizes the various approaches to fault prognosis in the literature.

Eventually, the model is allowed to perform long-term prognosis of the remaining useful life of the failing component/system with confidence bounds. The fault model PDF is convolved with the hazard zone PDF when the former reaches the threshold bounds and the resultant PDF is projected along the time axis (which is usually measured in "cycles" of operation) depicting the system's remaining life statistics.

Implementing physics-based models can provide a means to calculate the damage to critical components as a function of operating conditions and assess the cumulative effects in terms of component life usage. By integrating physical and stochastic modeling techniques, the model can be used to evaluate the distribution

Table 7.1 Approaches to fault prognosis in the engineering discipline

Source	Fault prognosis technique
Lewis [20]	Multistep adaptive Kalman filtering
Pham and Yang [31]	Auto-regressive moving-average models
Jardim-Gonçalves et al. [21]	Stochastic autoregressive integrated-moving—average models
Groer [22]	Weibull models
Frelicot [23]	Forecasting by pattern and cluster search
Ljung [24]	Parameter estimation methods
Aha [25]	Case-based reasoning
Studer and Masulli [26]	Petri nets, neural networks, fuzzy systems, and neuro-fuzzy systems
Tangirala [32]	Physics-based fatigue models

Sources found in Johnson [30]

of remaining useful life as a function of uncertainties in component strength/stress properties and loading conditions for a particular fault [33].

Bayesian estimation techniques can satisfy these requirements. In particular, particle filtering and learning strategies can be used for accurate and precise prediction of a failing component. This approach employs a state dynamic model and a measurement model to predict the posterior probability density function of the state to predict the time evolution of a fault or fatigue damage [6]. Unlike Kalman filtering particle filters allow a non-linear state dynamic model and non-Gaussian noise. Particle filters provide a robust framework for long-term prognosis while accounting effectively for uncertainties. Learning strategies can be applied to estimate correction terms to improve the accuracy and precision of the particle filtering algorithm for long-term prediction.

Particle filtering methods assume that the state equations that represent the evolution of the fault mode in time can be modeled as a first order Markov process with additive noise and conditionally independent outputs [6].

Let

$$x_k = f_{k-1}(x_{k-1}) + \omega_{k-1} \qquad (7.2)$$

$$z_k = h_k(x_k) + v_k \qquad (7.3)$$

where the state vector x_k includes a set of parameters that characterizes the evolution in time of the fault condition; the process noise ω_{k-1} represents the model uncertainty; z_k is the observation (measurement); and v_k is the measurement noise (uncertainty associated to sensors specifications and feature computation processes).

While there are several flavors of particle filters, the focus here is on algorithms based on the concept of *Sampling Importance Resampling* (SIR), in which the

7 Corrosion Diagnostic and Prognostic Technologies

posterior filtering distribution denoted as $\pi(x) = p(x_k|z_k)$ is approximated by a set of N weighted particles $\{\langle x_k^i, w_k^i \rangle; i = 1, \ldots, N\}$ sampled from an arbitrarily proposed distribution $q(x)$ that intends to be "similar" to $\pi(x)$ (i.e., $\pi(x) > 0 \Rightarrow q(x) > 0$ for all $x \in R^{n_x}$), The *importance weights* w_k^i are proportional to the likelihood $p(z_k|x_k^i)$ associated to the sample x_k^i, and normalized as in Eq. A25:

$$w_k^i = \frac{\pi(x_k^i)/q(x_k^i)}{\sum_{j=1}^{N} \pi(x_k^j)/q(x_k^j)} \tag{7.4}$$

such that $\sum_{j=1}^{N} w_k^i = 1$, and the posterior distribution (a.k.a the *target* distribution) can be approximated as

$$p(x_k|z_k) = \sum_{i=1}^{N} w_k^i \delta(x_k - x_k^i) \tag{7.5}$$

Thus, as in any Bayesian processor, the filtering stage is implemented in two steps: the computation of the a priori state density estimate (prediction step), and the update of the estimate according to the information presented by new measurements. Using the model in Eq. (7.3), the prediction step becomes

$$p(x_k|z_{k-1}) \approx \sum_{i=1}^{N} w_k^i f_{k-1}(x_k^i) \tag{7.6}$$

The update step, on the other hand, modifies the particle weights according to the relation

$$\bar{w}_k^i = w_k^i \frac{p(z_k|x_k^i) p(x_k^i|x_{k-1}^i)}{q(x_k^i|x_{k-1}^i, z_k)} \tag{7.7}$$

$$w_k^i = \frac{\bar{w}_k^i}{\sum_{j=1}^{N} \bar{w}_k^j} \tag{7.8}$$

It is possible that all but a few of the importance weights degenerate such that they are close to zero. In that case, one has a very poor representation of the system state (and wastes computing resources on unimportant calculations). To address that, *resampling* of the weights can be used [34]. The basic logical flowchart is shown in Fig. 7.25.

By using the state equation to represent the evolution of the fault dimension in time, it is possible to generate a long-term prediction for the state pdf, in the absence of new measurements, in a recursive manner using the current pdf estimate for the state,

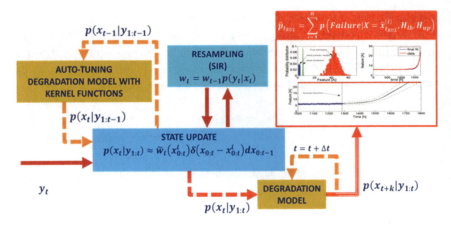

Fig. 7.25 The prognostic architecture using particle filtering

$$\tilde{p}(x_{t+p}|y_{1:t}) \approx \sum_{i=1}^{N} \tilde{p}(x_t|y_{1:t}) \prod_{j=t+1}^{t+p} p(x_j|x_{j-1}) dx_{1:t+p-1}, \quad (7.9)$$

which can be approximated as,

$$\tilde{p}(x_{t+p}|y_{1:t}) \approx \sum_{i=1}^{N} w_t^{(i)} \int \cdots \int p(x_{t+1:t+p-1}) \cdots \prod_{j=t+2}^{t+p} p(x_j|x_{j-1}) dx_{t+1:t+p-1} \quad (7.10)$$

The probability of failure at any future time instant is estimated by combining both the weights $w_{t+k}^{(i)}$ of predicted trajectories and specifications for the hazard zone through the application of the Law of Total Probabilities. The resulting RUL pdf, where t_{RUL} refers to RUL, provides the basis for the generation of confidence intervals and expectations for prognosis,

$$\tilde{p}_{t_{RUL}} = \sum_{i=1}^{N} p\left(Failure|X = \hat{x}_{t_{RUL}}^{(i)}, H_{lb}, H_{up}\right) \cdot w_{RUL}^{(i)}. \quad (7.11)$$

Figure 7.25 depicts the particle filtering module including the essential steps of degradation modeling, resampling, auto-tuning and state update.

Figure 7.26 illustrates the predicted fault growth of a system where a fault is detected at time t_{detect} and a prediction of the RUL is made at time $t_{prognosis}$. The probability of failure occurs outside the hazard-zone boundaries and is defined as the false-alarm rate α. The time corresponding to each predicted fault trajectory in the hazard-zone is represented as a distribution on the time-axis.. The maximum and minimum RUL values that encompass a confidence interval of value β are represented as t_{RUL}^{+} and t_{RUL}^{-}, accordingly.

7 Corrosion Diagnostic and Prognostic Technologies

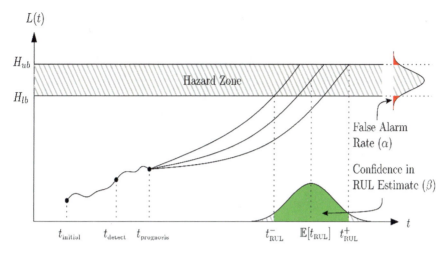

Fig. 7.26 Illustration of long-term prediction

The Georgia Tech research team pioneered the development and application of particle filtering methods to fault diagnosis and failure prognosis problems [6, 16, 35]. The prognostic framework is shown in Fig. 7.11. The fault is detected at t = 0. During the period from t = 0 to t = 1, data is streaming in and a feedback correction mechanism is used to update model parameters, thus reducing the inherent uncertainty. From $t = t_\theta$ on the model predicts the failing component's RUL. The threshold or hazard zone is empirically specified as a PDF. Model PDFs, at each step of the prognostic routine, are convolved with the hazard PDF and projected along the time axis to exhibit the statistics of the component's RUL.

These novel diagnostic and prognostic technologies have been applied to a variety of systems ranging from ground vehicles to rotorcraft, UAVs, and other industrial/military application domains. Of special note is the DARPA *Prognosis* program (Dr. George Vachtsevanos, PI) where this research team developed novel feature or condition indicator methods and diagnostic/prognostic algorithms. An experimental facility at Pax River was set up with a main transmission gearbox instrumented appropriately and installed on a NAVAIR rotorcraft. Tests were conducted (ground-air-ground) where a small initial carrier plate crack was allowed under nominal load conditions to grow up to a total plate failure. Blind tests carried out at Georgia Tech, as actual data were streaming in, demonstrated the efficacy of the prognostic approach. Predicted and actual crack growth data agreed well. Figure 7.27 shows the blind test results.

Figure 7.28 shows a GUI adopted for implementation, testing and demonstration of the overall health management framework.

Figure 7.29 is a flow chart for the particle filtering routine while Fig. 7.30 shows a flow chart for the prognostic algorithm.

- **Prognosis Case Study: Crack in Planetary Carrier Plate**

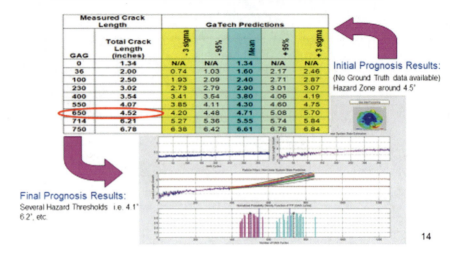

Fig. 7.27 Blind test results for carrier plate crack prediction (DARPA prognosis program)

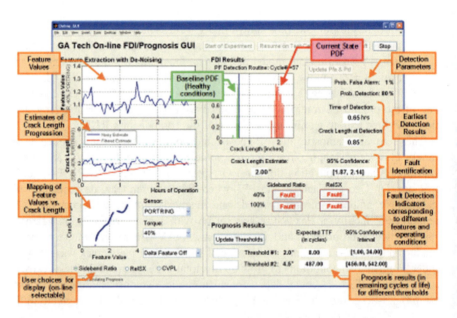

Fig. 7.28 The graphical user interface for the integrated health management framework

Prognosis, and thus the generation of long-term predictions, is a problem that goes beyond the scope of filtering applications because it involves future time horizons. Hence, a particle-filtering-based prognosis approach requires proposing a

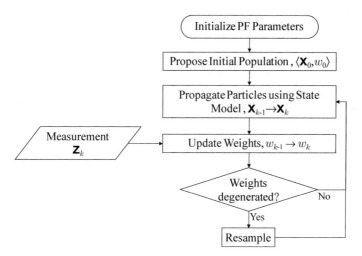

Fig. 7.29 Particle filtering flowchart

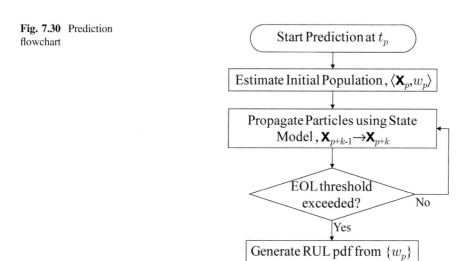

Fig. 7.30 Prediction flowchart

procedure to project the current estimate of the state probability density function (PDF) in time. The simplest implementation that can be used to solve this problem uses Eq. (7.1) recursively to propagate the posterior PDF estimate defined by $\{x_p^i, w_p^i;\ i = 1, \ldots, N\}$ in time, until x_p^i fails to meet the system specifications at time t_{EOL}^i. The RUL PDF—i.e., the distribution $p(t_{EOL}^i - t_p)$—is given by the distribution of w_p^i. Figure 7.26 shows the flow diagram of the prediction process.

7.19 Data-Driven Prognostic Techniques

In some cases involving complex systems, it may be difficult or impossible to derive dynamic models based on all the physical processes involved. In such cases, it is possible to assume a certain form for the dynamic model and then use observed inputs and outputs of the system to determine the model parameters needed so that the model indeed serves as an accurate surrogate for the system. This is known as model identification.

For fault diagnosis and failure prognosis, a variety of input-output mappings have been employed as surrogate system models. Specifically, one may have historical fault/failure data in terms of time plots of various signals leading up to failure, or statistical data sets. In such cases, it is very difficult to determine any sort of model for prediction purposes. In such situations, one may use nonlinear network approximations that can be tuned using well-established formal algorithms to provide desired outputs directly in terms of the data. They provide structured nonlinear function mappings with very desirable properties between available data and desired outputs.

The book Applications of Intelligent Control to Engineering Systems [36] presents alternative tools for both forecasting researchers and practitioners. Namely, these tools include artificial neural networks (ANNs), fuzzy systems, and other computational intelligence methods based on the linguistic and reasoning abilities of humans. It has been reported that ANNs trained with the backpropagation algorithm outperform traditional statistical methods such as regression and Box-Jenkins approaches [36]. ANNs are data-driven, self-adaptive, and they make very few assumptions about the models for problems under study. The aim of ANNs is to learn subtle functional relationships among the data from example data. Thus, ANNs are well suited for practical problems, where there is an abundance of data and a lack of knowledge of the underlying system's fault behavior. ANNs can be viewed as one of many multivariate nonlinear and nonparametric statistical methods [36]. Data-driven approaches to failure prognosis also take advantage of recurrent neural networks, dynamic wavelet neural networks, neuro-fuzzy systems, and a variety of statistical tools. For training and validation, they use the current and past history of input data and feedback outputs via unit delay lines. The main problem of ANNs is that their decisions are not always evident and they can be over-trained. Nevertheless, it has been shown that ANNS provide a feasible tool for practical prediction problems [36].

Model-Based Prognostic Techniques

Model-based prognostic schemes include those that employ a dynamic model of the process being predicted. These can include physics-based models, autoregressive moving-average (ARMA) techniques, Bayesian filtering algorithms, and empirical-based methods. Model-based methods provide a technically comprehensive approach that has been used traditionally to understand component failure mode progression.

Figure 7.31 depicts a model-based prognostic scheme. Input from the diagnostic block is combined with stress profiles and feeds into the fault growth model. An estimation method (in this case particle filtering) is called upon to propagate the fault model initially one step at a time while model parameters are updated on-line in real-time as new sensor data become available.

Eventually, the model is allowed to perform long-term prognosis of the remaining useful life of the failing component with confidence bounds. The fault model PDF is convolved with the hazard zone PDF when the former reaches the threshold bounds and the resultant PDF is projected along the time axis depicting the component's remaining life statistics.

The designer of a comprehensive and verifiable prognostics architecture is faced with significant challenges: data availability (baseline and fault data that is correlated and time stamped, sampled at appropriate rates) is always a concern; uncertainty representation, propagation and management, inherent in prognosis; high fidelity modeling of critical components/systems is lacking; data mining tools for feature extraction and selection are hand-engineered; finally, diagnostic and prognostic algorithms must address accurately the designer's specifications. There is a need to differentiate between *prognosis* and *trending*, i.e. the practice of regressing linearly a system variable (temperature, for example) until it reaches a specified threshold. Another differentiating characteristic relates to health-based vs usage-based prognostics with the former predicting the RUL of failing components on-line in real time as data is streaming in after a fault or incipient failure has been detected and identified. The latter refers to long-term prediction of the RUL exploited for reliability and life cycle management studies. We address both prognostic notions in this paper and view each one within the overall context and scope of the health-based vs usage-based framework. The theoretical underpinnings borrow from the emerging fields of Prognostics and Health Management (PHM),

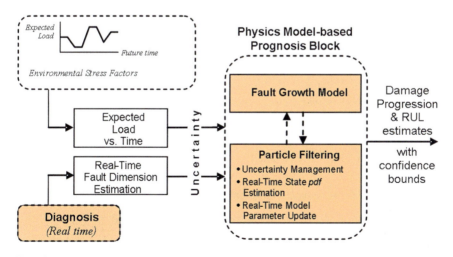

Fig. 7.31 A scheme for model-based prognosis

Condition Based Maintenance (CBM+) and novel data mining, reasoning, and work package optimization to expedite maintenance actions at the depot level. The enabling technologies build on mathematically rigorous concepts from Bayesian estimation theory—an emerging and powerful methodology called Particle Filtering (PF) that is particularly useful in dealing with difficult non-linear and/or non-Gaussian problems; high fidelity modeling of critical components/systems and novel data mining tools augmented with Deep Learning methods for optimum feature selection and extraction; a reasoning paradigm, called Dynamic Case Based Reasoning (DCBR), is exploited as the dynamic knowledge base entailing attributes of learning and adaptation; optimized maintenance planning and scheduling aimed to expedite the interface between on-platform events (faults/failures) and depot maintenance practices. All algorithmic developments are accompanied by appropriate performance and effectiveness metrics to support development and validation tasks. Moreover, the enabling technologies take advantage of Model Predictive Control and fault adaptive control methods to execute control reconfiguration, redistribution and mission adaptation in order to safeguard the integrity of the vehicle for the duration of an emergency. We introduce novel tools/methods to represent and manage uncertainty, inherent in prognostics.

Prognosis is activated when a corrosion degradation/fault is detected. For the same degradation mode, the propagation of the fault follows the same physical law. Therefore, in the prognosis model, the Boolean state can be removed from the model (1) and we get a model for prognosis as follows:

$$\begin{cases} x_c(t+1) = f_t(x_c(t), \omega(t)) \\ \text{Features}(t) = h_t(x_c(t), v(t)) \end{cases} \quad (7.12)$$

The definitions of symbols are the same as in (1). Again, we will only focus on the equation mainly describes the progression of the fault. It is given as:

$$x_c(t+1) = (1+\beta)x_c(t) + \omega(t) \quad (7.13)$$

Note that Eq. (7.3) is a special case of the second equation in model (7.2). In the second equation in model (7.2), $x_{d,2} = 1$ for a faulty condition while $x_{d,2} = 0$ for a healthy condition. When a fault is detected, $x_{d,2} = 1$ and, therefore, they are exactly the same for a faulty condition.

7.20 Model On-Line Update

Figure 7.32 is a schematic representation of the model parameter update framework.

The propagation of the corrosion state under tightly controlled conditions could show significantly different behaviors. Therefore, the deterministic model must be modified to take into consideration the uncertainty due to the stochastic

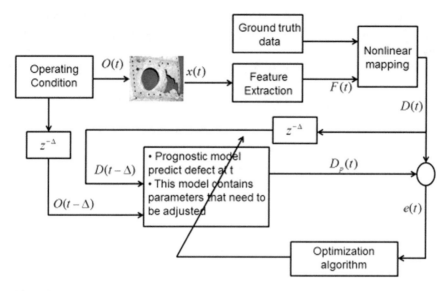

Fig. 7.32 A schematic representation of the model parameter update configuration

characteristics of the degradation model by adding a random variable. In practice, adding a random variable into the growth law is the same as adding a random variable into its parameters, and we arrive at

$$L(t+\Delta t) = L(t) + \Delta t C_L (L(t))^m$$
$$C_L = C_L + \omega_{C_L} \quad (7.14)$$
$$m = m + \omega_m$$

where C_L and m can be regarded as states associated with the model, ω_{C_L} and ω_m are zero mean random noise.

With unit step size, it can be modified as

$$L(t+1) = L(t) + p_1(t) C_L (L(t))^{p_2(t)m}$$
$$C_L = C_L + \omega_{C_L} \quad (7.15)$$
$$m = m + \omega_m$$

Thus, two parameters $p_1(t)$ and $p_2(t)$ are introduced to facilitate the online parameter adaptation scheme.

To determine the parameters, a recursive least squares algorithm with a forgetting factor is employed since it is generally fast in its convergence.

Thus, two parameters $p_1(t)$ and $p_2(t)$ are introduced to facilitate the online parameter adaptation scheme.

To determine the parameters, a recursive least square algorithm with a forgetting factor is employed since it is generally fast in its convergence. The algorithm is implemented as follows:

Step 1: define a cost function as:

$$J(\theta) = \frac{1}{2}\sum_{t=1}^{T}\lambda^{T-t}\left[D(t) - D\left(\hat{\theta}(t-1)\right)\right]^2 \quad (7.16)$$

where λ is the forgetting factor, which is usually given in the range of $0 < \lambda \leq 1$, and $\theta = [p_1(t) p_2(t)]^T$ is the parameter vector to be determined.

Step 2: Calculate the derivatives with respect to parameters Θ:

$$\phi(t) = \frac{d\hat{D}(t,\theta)}{d\theta} \quad (7.17)$$

Step 3: The parameter update is given by:

$$\hat{\theta}(t) = \hat{\theta}(t-1) + P(t)\phi(t)\left[D(t) - D\left(\hat{\theta}(t-1)\right)\right] \quad (7.18)$$

and $P(t)$ is updated as

$$P(t) = \frac{P(t-1)}{\lambda}\left[1 - \frac{\phi(t)\phi^T(t)P(t-1)}{\lambda + \phi^T(t)P(t-1)\phi(t)}\right] \quad (7.19)$$

The recursive least square with a forgetting factor actually applies an exponential weighting to the past data. In the cost function (A9), the influence of past data reduces gradually as new data become available. This algorithm can be easily applied on-line.

To implement the algorithm, a set of initial parameters must be given. Parameter $\theta(0)$ is given according to our prior knowledge of the system while $P(0)$ is given as a large number times an identity matrix.

A flow chart of this on-line adaptation algorithm is given in Fig. 7.33. The first step is to initialize the parameters used in the model. Currently, the initial value is obtained via a priori knowledge of the system. With more experimental data, learning algorithms, such as neural networks, can be implemented to train the model so that its initial value is close to the actual one. The on-line adaptation routine will be more efficient with good initial parameters.

Note that the parameter adaptation is realized by a recursive least square method. Other methods, such as an extended Kalman filter or a neural network, etc., can be used as well.

7 Corrosion Diagnostic and Prognostic Technologies

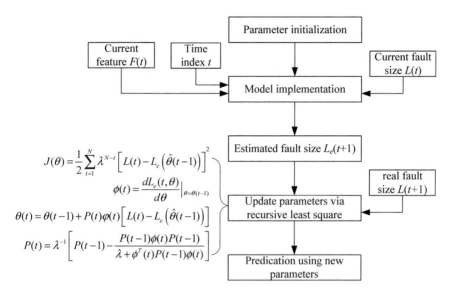

Fig. 7.33 Flow chart of the model on-line update

7.21 Consideration of Operating Conditions

In the previous model, the operating conditions such as ambient temperature, humidity, PH, etc., are not taken into consideration. The influence of the operating and environmental conditions is reflected by the on-line parameter tuning. Moreover, stress conditions must be accounted for when appropriate. If the operating conditions can be compensated, a precise fault propagation model can be derived.

Let us consider the corrosion fault mode. It is known that humidity is a leading contributing factor of corrosion. Therefore, the environmental humidity should be incorporated in the corrosion propagation model. Suppose that the nominal humidity (could be normal room humidity) is denoted by H_n. The environmental humidity is measured as H_c. The normalized humidity condition then can be described as a humidity index h_i which is given as $h_i = H_c/H_n$. Clearly, large humidity values result in larger h_i, while small humidity values result in smaller h_i.

If we know that the humidity influences linearly the corrosion propagation, the previous modified equation for the degradation rate can be further re-written as (7.20) to include the humidity factor.

$$\dot{D} = \frac{dD}{dt} = h_i C_D (D)^n \qquad (7.20)$$

If, however, we know that humidity influences exponentially the corrosion propagation, (7.20) can be re-written as:

$$\dot{D} = \frac{dD}{dt} = C_D(D)^{n \cdot h_i} \tag{7.21}$$

Accordingly, the discrete time form model should be modified as:

$$D(t+1) = D(t) + h_i C_D(D(t))^n, \tag{7.22}$$

and

$$D(t+1) = D(t) + C_D(D(t))^{n \cdot h_i}, \tag{7.23}$$

respectively.

Relative humidity may also be included as a time-varying parameter in the model, just as other model parameters. When other operating conditions are included, such as nominal temperature T_n (normal operating temperature under healthy conditions), can be defined as well. Then, a temperature index t_i is assigned to represent this condition. Multiple ways are available to combine them into a single model.

Suppose these factors influence linearly the fault propagation, a possible alternative is to write the model as either

$$D(t+1) = D(t) + h_i t_i l_i g_i C_D(D(t))^n, \tag{7.24}$$

or

$$D(t+1) = D(t) + \left(w_h h_i + w_t t_i + w_l l_i + w_g g_i\right) C_D(D(t))^n, \tag{7.25}$$

where w_h, w_t, w_l, w_g are weighting factors and $w_h + w_t + w_l + w_g = 1$.

The model can be written as either

$$D(t+1) = D(t) + C_D(D(t))^{n h_i t_i l_i g_i} \tag{7.26}$$

or

$$D(t+1) = D(t) + C_D(D(t))^{n\left(w_h h_i + w_t t_i + w_l l_i + w_g g_i\right)}, \tag{7.27}$$

when the factors are exhibiting an exponential dependence.

It is possible that the influence of some factors is exhibiting a linear behavior, while that of others is exponential. In this case, suppose the linear factors are $f_{l,1}$ and $f_{l,2}$ and the exponential factors are $f_{e,1}$ and $f_{e,2}$. Then, the model can be written as

$$D(t+1) = D(t) + f_{l,1} f_{l,2} C_D(D(t))^{n f_{e,1} f_{e,2}}, \tag{7.28}$$

or

$$D(t+1) = D(t) + (w_{l,1}f_{l,1} + w_{l,2}f_{l,2})C_D(D(t))^{n(w_{e,1}f_{e,1} + w_{e,2}f_{e,2})}, \qquad (7.29)$$

where $w_{1,1}$, $w_{1,2}$, $w_{e,1}$, $w_{e,2}$ are weighting factors and $w_{1,1} + w_{1,2} = 1$ and $w_{e,1} + w_{e,2} = 1$.

7.22 Statistical Techniques

For situations in which sophisticated prognostic models are not or cannot be utilized (perhaps due to disadvantageous ROI, low failure-criticality rates, or where there is an insufficient sensor network to assess condition), a statistical reliability or usage-based prognostic approach may be the only alternative. This form of prognostic algorithm is the least complex and requires that a history of component failure or operational usage profile data is available. One typical approach would be to fit a Weibull distribution (or other statistical failure distribution) to such failure or inspection data [22, 37]. Despite the obvious loss of information (compared to condition-based approaches), a statistical reliability-based prognostic distribution can still be used to drive interval-based maintenance practices which can then be potentially revised by the information obtained from maintenance. The benefit of a regularly updated maintenance database is critical for this approach.

7.23 Particle Filtering as an Uncertainty Representation and Management Technique for Failure Prognosis

Uncertainty in prognosis is probably the most significant challenge facing the PHM system designer. Uncertainty management tools seek to improve the fault signal to noise (uncertainty) ratio. They begin by determining the uncertainty sources in terms of an uncertainty tree and then exploit filtering or kernel-based methods for uncertainty management [18, 38]. We use particle filtering, as a nonlinear filtering method employing noisy observation data to estimate at least the first two moments of a state vector governed by a dynamic nonlinear, non-Gaussian state-space model. From a Bayesian standpoint, a nonlinear filtering procedure intends to generate an estimate of the posterior probability density function $p(x_t|y_{1:t})$ for the state, based on the set of received measurements. Particle Filtering intends to solve this estimation problem by efficiently selecting a set of N particles $\{x^{(i)}\}_{i=1\cdots N}$ and weights $\{w_t^{(i)}\}_{i=1\cdots N}$, such that the state pdf may be approximated by

$$\tilde{\pi}_t^N(x_t) = \sum_{i=1}^{N} w_t^{(i)} \delta(x_t - x_t^{(i)}).$$

The evolution in time of the fault dimension is expressed through the nonlinear state equation:

$$\begin{cases} x_1(t+1) = x_1(t) + x_2(t) \cdot F(x(t), t, U) + \omega_1(t) \\ x_2(t+1) = x_2(t) + \omega_2(t) \end{cases},$$

where $x_1(t)$ is a state representing the fault dimension under analysis, $x_2(t)$ is a state associated with an unknown model parameter, U are external inputs to the system (load profile, etc.), $F(x(t), t, U)$ is a general time-varying nonlinear function, and $\omega_1(t)$, $\omega_2(t)$ are white noises (not necessarily Gaussian). The nonlinear function $F(x(t), t, U)$ may represent a model based on first principles, a neural network, or even a fuzzy system.

Consider a discrete approximation for the predicted state pdf:

$$\hat{p}(x_{t+k}|\hat{x}_{1:t+k-1}) \approx \sum_{i=1}^{N} w_{t+k-1}^{(i)} K\left(x_{t+k} - E\left[x_{t+k}^{(i)}|\hat{x}_{t+k-1}^{(i)}\right]\right),$$

where $K(\cdot)$ is a kernel density function, which may correspond to the process noise pdf, a Gaussian kernel or a rescaled version of the Epanechnikov kernel:

$$K_{opt}(x) = \begin{cases} \frac{n_x+2}{2c_{n_x}} \left(1 - \|x\|^2\right) & \text{if } \|x\| < 1 \\ 0 & \text{otherwise} \end{cases},$$

where c_{n_x} is the volume of the unit sphere. Furthermore, if the density is Gaussian with unit covariance matrix, the optimal bandwidth is given by

$$h_{opt} = A \cdot N^{-\frac{1}{n_x+4}}$$

$$A = \left(8 c_{n_x}^{-1} \cdot (n_x + 4) \cdot (2\sqrt{\pi})^{n_x}\right)^{\frac{1}{n_x+4}}.$$

The Epanechnikov kernel is well suited for uncertainty representation in long-term prediction. Given $\{x_t^{(i)}\}_{i=1\cdots N}$ and $\{w_t^{(i)}\}_{i=1\cdots N}$ as initial conditions, it is possible to represent the uncertainty inherent to the predicted state pdf by performing an inverse transform resampling procedure for the particle population [12]. This method obtains a fixed number of samples for each future time instant, avoiding problems of excessive computational effort. In fact, after the resampling scheme is performed, the weights may be expressed as: $\{w_{t+k}^{(i)}\}_{i=1\cdots N} = N^{-1}$. Furthermore, if only Epanechnikov kernels are used, it is ensured that the

Fig. 7.34 Particle filtering-based uncertainty representation; results for the case of a fatigue fault progression in a critical aircraft component

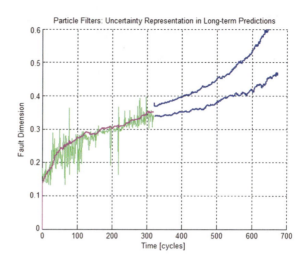

representation of the uncertainty will be bounded. These bounds intrinsically incorporate, measure, and represent model uncertainty (through the estimation of unknown parameters) and measurement noise (since the initial condition for long-term predictions corresponds to the output of the particle filtering procedure).

For the test case of a fatigue failure in a critical aircraft component, the uncertainty representation results are shown in Fig. 7.34.

7.24 Uncertainty Management in Long-Term Predictions

The issue of uncertainty management, in a Particle Filtering-based prognosis framework, is related to a set of techniques aimed to improve the estimate at the current time instant, since both the expectation of the predicted trajectories for particles and bandwidth of Epanechnikov kernels depend on that pdf estimate.

In this sense, it is important to distinguish between two main types of adjustments that may be implemented to improve the current representation of uncertainty for future time instants:

- Adjustments in unknown parameters in the state equation.
- Adjustments in the parameters that define the noise pdf embedded in the state equation, known as "hyper-parameters".

Outer correction loops may be also implemented using neural networks, fuzzy expert systems, PID controllers, among others. Additional correction loops include the modification of the number of particles used for 1-step or long-term prediction purposes.

7.25 Measuring Prognostics Performance

A number of measures that can be used to evaluate prognostic performance have been proposed in the recent past. Some metrics of particular interest are listed below.

7.26 Prognostic Horizon

Prognostic Horizon (PH) fulfills two roles: first, it identifies whether an algorithm predicts within a specified error margin (specified by α, a statistical confidence parameter) around the actual End-of-Life (EOL, the time index for actual end of life, according to the defined failure threshold); *and, second, it indicates how much time the algorithm provides for any corrective action to be taken.* In other words, it assesses whether an algorithm yields a sufficient prognostic horizon; if it does not, it may not be useful or meaningful to compute other metrics. PH is defined as the difference between the EoP (End-of-Prediction) and the current time index i, utilizing data accumulated up to the time index i and provided the prediction meets desired specifications. This specification may be defined in terms of an allowable error bound (α) around true EOL. It is expected that PHs are determined for an algorithm-application pair offline during the validation phase. These numbers can then be used as guidelines when the algorithm is deployed in test applications when the actual EOL is not known in advance. While comparing algorithms, an algorithm with a longer prediction horizon H would be preferred. The prediction horizon is computed as $H = EoP - i$, where $i = min\{j | (j \in l) \wedge (r_*(1 - \alpha) \leq r^l(j) \leq r_*(1 + \alpha))\}$, $r_*^l(i)$ is the true RUL at time i given that data is available up to time i for the lth UUT (unit under test), and $r^l(i)$ is the RUL estimate for the lth UUT at time i as determined from measurement and analysis.

For instance, a PH with error bound of $\alpha = 0.05$ identifies when a given algorithm starts predicting estimates that are within 5% of the actual EOL. Other specifications may be used to derive PH as desired.

7.27 α-λ Performance

α-λ performance identifies whether the algorithm performs within desired error margins (specified by the parameter α) of the actual RUL at any given time instant (specified by the parameter λ) that may be of interest to a particular application. This presents a more stringent requirement of staying within a converging cone of error margin as a system nears EOL. The time instances may be specified as percentage of total remaining life from the point the first prediction is made or a given absolute time interval before EOL is reached. For instance, we define α-λ

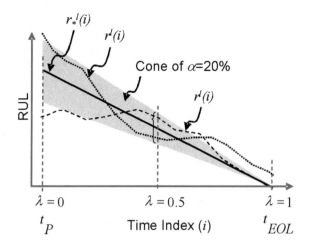

Fig. 7.35 α-λ performance visualization

accuracy as the prediction accuracy to be within $\alpha \cdot 100\%$ of the actual RUL at specific time instance t_λ expressed as a fraction of time between the point when an algorithm starts predicting and the actual failure. For example, this metric determines whether a prediction falls within 20% accuracy (i.e., $\alpha = 0.2$) halfway to failure from the time the first prediction is made (i.e., $\lambda = 0.5$). The metric is visualized in Fig. 7.35. An extension of this metric based on other performance measures is straightforward:

$$[1-\alpha]r_*(t) \leq r^l(t_\lambda) \leq [1+\alpha]r_*(t) \qquad (A30)$$

where α is the accuracy modifier, λ is the time window modifier, $t_\lambda = P + \lambda(EOL - P)$, and P is the time index at which the first prediction is made by the prognostic system.

Note that α-λ Performance and Prognostic Horizon can also be computed as precision metrics.

7.28 Prognostic Dynamic Standard Deviation (DSTD)

Other performance measures intend to quantify the volatility of generated predictions, something that can be achieved by computing the standard deviation of the expected EOL over a sliding window:

$$DSTD = \varphi\left(\sqrt{Var\left(E\{EOL|y_{1:j}\}\right)_{j=k_{pred}-\Delta:k_{pred}}}\right)_{\forall k_{pred}\in[1,EOL]} \qquad (7.30)$$

where k_{pred} is the cycle in which the prognostic algorithm is executed, Δ is the number of samples considered in the sliding window, φ is the logistic function that aims to scale the results in the range [0,1]. For this measure, the better DSTD, the closer to zero is the measure (*DSTD* = 0 indicates perfect null volatility and the fact that new measurements do not alter the output of the prognostic algorithm).

7.29 Critical-α Index

Decision-making support-systems cannot depend solely on information about the expectation of random variables, because the tails of PDFs contain critical information about the risk that is associated to process operation. The *critical-α* index is a measure based on the concept of the JITP (Just in Time Point) that helps to quantify this point. The *critical–α* index is a measure of risk aversion (a significant factor to be considered when using implementations that overestimate the remaining useful life of a system) and is defined as the maximum $\alpha \in [0,100]$ that guarantees that the JITP$_{\alpha\%}(k_{pred})$ value is smaller than the ground truth value of the EOL time instant, for all $k_{pred} \in [1, EOL]$:

$$\alpha_{crit} = \arg\max_{\alpha}\left\{ JITP_{\alpha\%}(k_{pred}) \leq EOL \right\}_{\forall k_{pred} \in [1, EOL]} \quad (7.31)$$

Decision-making support systems that consider prognostic algorithms with larger critical–α values in their design are capable of implementing strategies that are more aggressive. This is because these prognostic routines are conservative; thus, it is possible to accept the risk of accumulating larger failure probability mass before recommending a corrective action. However, a large critical–α value is also an indicator that the variance of the predicted EOL PDF is large (i.e., less precise estimates of the EOL). For this reason, a good design should try to lessen this problem by selecting prognostic algorithms that allow not only the use of large critical-α values, but also minimize—over time—the difference between the ground truth EOL and the JITP values computed for the corresponding $\alpha_{crit}\%$.

Performance and Effectiveness Metrics Performance and effectiveness metrics are the absolute pre-requisites for defining and specifying requirements in the development of diagnostic and prognostic algorithms. They drive the derivation of accurate and reliable requirements and specifications in the systems engineering process. Fault diagnosis metrics, such as confidence (accuracy) and false alarm rates can be defined and determined numerically or in statistical/probabilistic terms. Accuracy and precision cannot be determined while we are predicting the time to failure or remaining useful life of a failing component since, for causal systems, we do not know the true failure time a priori. We can only observe the trending of the fault evolution in real time and infer appropriate metrics from these observations.

Diagnostics Performance Metrics Confidence (accuracy), False Alarm Rate (False Positives/False Negatives), Response to Uncertainty/Noise, Computational Complexity, Time to detection from the inception of a fault—important for timely prognosis (or Time delay—time span between the initiation and the detection/isolation/identification of a fault/failure event), Ability to perform on-line in real-time, a host of other metrics for fault diagnosis relate to coverage (how many critical faults can be detected), isolability (ability to isolate down to the LRU level), detectability (ability to distinguish between multiple fault modes), etc.

Prognostics Performance Metrics [39, 40] We list four metrics that can be used to evaluate prognostic performance. These four metrics follow a systematic progression in terms of the information they seek. The first metric called Prognostic Horizon (PH) identifies whether an algorithm predicts within a specified error margin around the actual EoL and if it does how much time it allows for any corrective action to be taken. If an algorithm passes the PH test, the next metric called α-λ which identify whether the algorithm performs within desired error margins (specified by the parameter α) of the actual RUL at any given time instant (specified by the parameter λ). If this criterion is also met, the next step is to quantify the accuracy levels relative to actual RUL. This is accomplished by the metrics called Relative Accuracy (RA) and Cumulative Relative Accuracy (CRA). These notions assume that prognostics performance improves as more information becomes available with time and hence by design an algorithm will satisfy these metrics criteria if it converges to true RULs. Therefore, the fourth metric Convergence quantifies how fast the algorithm converges if it does satisfy all the previous metrics. These metrics can be considered as a hierarchical test that provides several levels of comparison among different algorithms in addition to the specific information the metrics individually provide regarding algorithm performance (Figs. 7.36, 7.37 and 7.38).

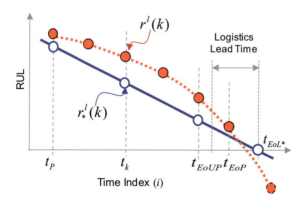

Fig. 7.36 An illustration depicting some important prognostic time definitions and concepts

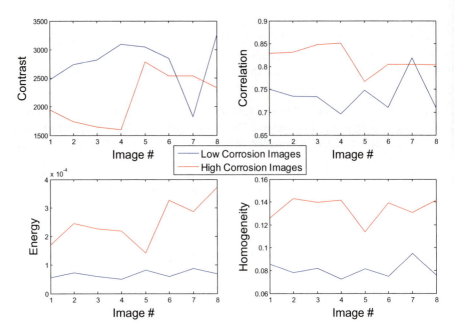

Fig. 7.37 Contrast, correlation, energy and homogeneity features of low and high corrosion images (image number ascends correspond to the sequence from left to right)

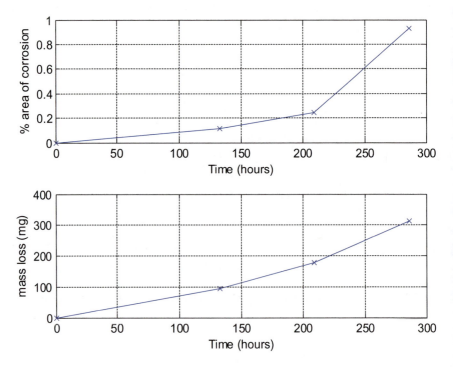

Fig. 7.38 Top: Corroded area percentage over time. Bottom: Measured mass loss (mg) over time

7 Corrosion Diagnostic and Prognostic Technologies

7.30 Corrosion Diagnostic and Prognostic Results

The diagnostic and prognostic technologies suggested above were applied, in simulation, to the test panel corrosion surfaces. The simulation, conducted in Simulink, utilizes the pitting model in Eq. (7.6) and the cracking progression model (Paris' Law) in Eq. (7.9). Figure 7.39 shows the results of the particle filtering diagnostic routine on the pit depth calculated from μLPR data (Table 7.2).

The top graph shows the calculated pit depth (green) and the particle filter based estimate of the pit depth (black). The bottom graph shows the detection confidence at each time step. The blue histogram is the baseline distribution and the red one is the current distribution. The black line is the threshold which set the Type I error as 5%. From the figures, it is shown that from time 18 h, when enough data is collected for generating baseline distribution, to time 36 h, Type II error is reduced from 78 to 2%, which means the confidence of abnormal is increased from 22 to 98%. So, at time 36 h, abnormal is declared.

After the diagnostic routine detects corrosion to a specified confidence level, the prognosis routine is run, as shown in Fig. 7.40. The dashed blue line is the measured data of the fault level up to the time of detection. Once corrosion is detected the particles are allowed to progress using the non-linear state model. The failure threshold is user specified and is given as a PDF. The particles progression and the failure threshold PDF is used to calculate the time to failure PDF. The empirical model from Eq. (7.6) can be employed where $b = 1.54$. The following figure is the result of model tuning and prediction. Time is divided into three sections, which are

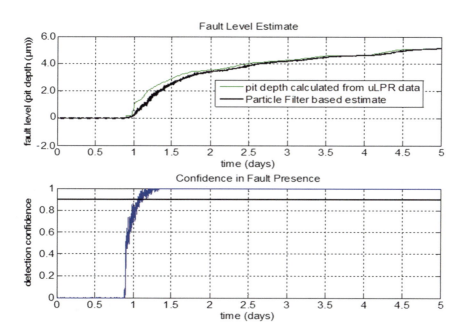

Fig. 7.39 Pit depth from μ LPR data using the particle filtering methodology

separated by two dashed lines at time 36 h and 270 h, respectively. Initially, the model parameters are tuned on the basis of available data, as described above. The blue line and green line are upper bound and lower bound of pit depth, respectively. The red dots are the measured pit depth. After the model is tuned, the prediction routine is initiated using only the model. Threshold is shown as the red line, which is 10 μm. In this case, the predicted remaining useful life is shown in Fig. 7.41 as a histogram or approximate PDF. The prognostic results are matching well-anticipated corrosion predictions in real on-board aircraft applications. The Air Force project reported in this paper has not proceeded to its final stage of on-aircraft testing, data collection, testing and evaluation of diagnostic and prognostic algorithms. Simulation studies and results show the efficacy of the integrated approach to testing, data mining, corrosion initiation and prediction of global and localized corrosion processes.

Table 7.2 FDR values of image features

Features	Contrast	Correlation	Energy	Homogeneity
FDR	0.9604	2.2084	95.1962	27.3738

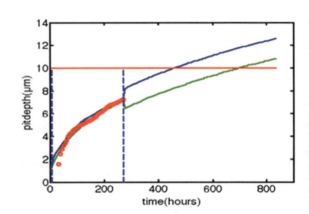

Fig. 7.40 Pit depth prediction scheme

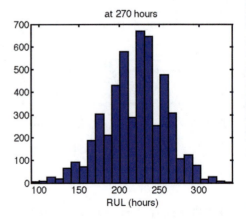

Fig. 7.41 The remaining useful life prediction at 270 h

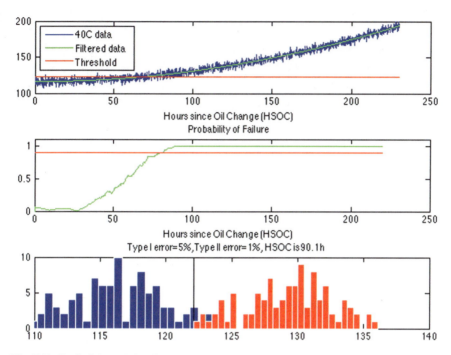

Fig. 7.42 Typical degradation detection results

The baseline (no degradation) result is generated from the first three data points. We use this to fit a normal distribution and then derive samples from this distribution.

A typical detection result is shown in Fig. 7.42. Step 2, i.e. prediction, is triggered after the detection step is completed before the dashed line. Prognosis is triggered after 90 h.

7.31 Testing of Data-Mining Techniques on μLPR Sensor Data

Kolmogorov Complexity (KC) is a measure of randomness of data based on their information content. The KC anomaly detection algorithm utilizes the Compression-based Dissimilarity Measure (CDM) to recursively determine which half of the data is more complex. The CDM is given by the following equation: $CDM(x,y) = \frac{C(xy)}{C(x)+C(y)}$ where C(x) approximates the Kolmogorov Complexity of x. The bottom graph in Fig. 7.1 shows the CDM of the LPR data (top Fig. 7.43) and the baseline data (μLPR data expected when there is zero corrosion) as function of time.

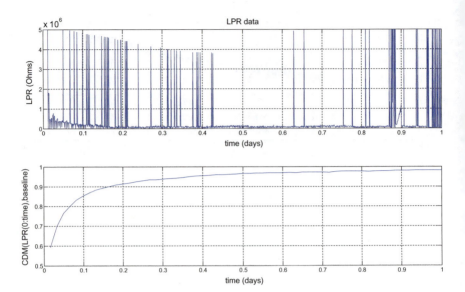

Fig. 7.43 CDM of LPR Data

Imaging Techniques for Corrosion Extent Determination

High-resolution images of coupons are used as a reference measure of corrosion for μLPR experiments. The imaging technique consists of the following steps:

1. Threshold.

    (a) Threshold image into regions of varying levels of corrosion. For example threshold image into regions of High corrosion, Medium corrosion and No corrosion.

    (b) Figures 7.44 and 7.45 show an example of using a 5 layer density threshold. Figure 7.44 shows the original image and Fig. 7.45 shows the threshold image. The threshold levels were determined using minimum a variance quantization method.

Fig. 7.44 Original image

Fig. 7.45 Threshold image

2. Corrosion extent determination.

   a. Determine the degree of corrosion of the threshold levels.

   - This can be based on color, pixel range, smoothness, contrast etc.

   b. The extent of corrosion can be determined by measuring the threshold area against the full image size.

- *Particle Filtering*

Figure 7.46 shows the particle filter estimate and detection confidence generated by using the particle filtering routine on synthetic data. The particle filtering diagnostic method is outlined below:

1. Initialize particle filter

   - Generate probability density function (PDF) from baseline data
   - From baseline data generate initial particles

2. Run particle filter

   - Update particles
   - Update weights
   - Get current PDF from particles and weights
   - Determine detection confidence

The Association Strategy—Relating Sensor Outputs (Features) to Control Decisions

Various algorithms are available to carry out efficiently and effectively the classification task. We will explore an "intelligent" method to map sensor outputs in the form of features to action decisions. The proposed approach will rely on a Dynamic Bayesian Network (DBN) as the association paradigm. Dynamic Bayesian Networks (DBNs) are developed for explicating the dynamic cause-effect relationships with the strength of the causal connection amongst the problem situations. DBNs are inference networks that use cyclic, directed graphs representing the

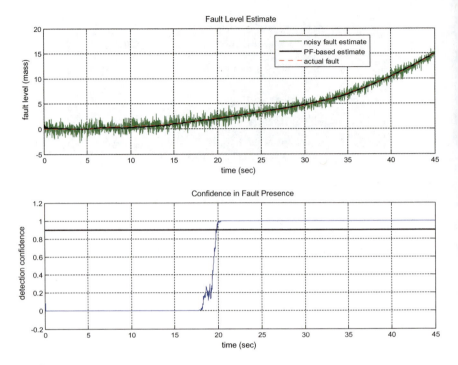

Fig. 7.46 Particle filtering diagnostics example

causal relationships between concepts. DBNs can easily represent structured knowledge, especially for those containing causal relationships compared with neural networks. Each DBN represents the knowledge of an expert. Several DBNs can be combined into one by merging their adjacency matrices with different, weighted coefficients that usually represent beliefs of different human experts.

The key to stochastic modeling is to process data/information that is directly relevant to the goals at hand.

- Perception. This is the most primitive level of situational awareness. Providing more data of better accuracy in shortened times can improve level 1 situational awareness.
- Comprehension. In level 2 situation awareness, a decision maker integrates a variety of data with the goals and constraints they are operating under to develop an assessment of the situation.
- Projection. This highest level of situation awareness determines the best or most appropriate response by predicting the resulting courses of actions.

7.32 Performance Evaluation

We will establish and exercise appropriate performance metrics at each level of the proposed learning and decision support hierarchy.

The DCBR performance will be evaluated on the basis of the following criteria:

- Total solution search time—total time to search and rank matching cases.
- Quality of solution—relevance of retrieved cases; success and failure record; success of the top ranked case; overall success if several iterations are required.
- Comparison to legacy systems.

With regard to the incremental learning process, the emphasis, from a performance evaluation perspective, is on *correctness*. We will define appropriate correctness, precision and recall metrics for this purpose.

7.33 Performance Metrics/Specifications/Constraints

We will be defining generic performance metrics and specifications such as:

- Computational Speed—Data processing and image analysis algorithms must be executed in an expedient manner. It is crucial that performance metrics be defined and requirements specified on computational speed. Such metrics will dictate necessary tradeoffs between accuracy and algorithm sophistication and execution time.
- Accuracy—Accuracy metrics are defined relating to data mining, corrosion detection and prediction.
- Precision—The ability to pinpoint and resolve the existence of degrading conditions.

Impact of Corrosion Induced Fatigue on the Integrity of Critical Assets

The potential impact of corrosion on the asset's integrity may be assessed now providing accurate and timely information to the maintainer when remedial action must be taken. The key "smart" sensor ingredients are in place:

- Damage detection provides a trigger for maintenance action.
- Knowledge of damage and corrosion tolerance of the structure in adverse environments.
- Parametric modeling allows for inclusion of stress crack profiles.
- Low resistance and spikes in the LPR measurements are indicative of pitting corrosion.

Figure 7.47 depicts the overall architecture for corrosion assessment.

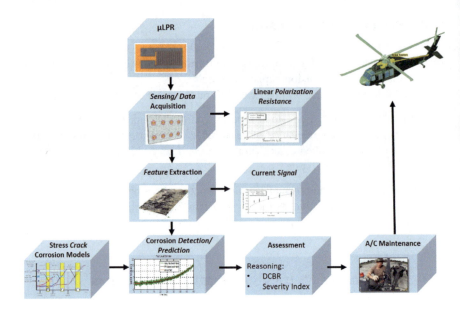

Fig. 7.47 A framework to assess the impact of corrosion on the integrity of the asset

Corrosion Prediction in Aging Aircraft Materials Using Neural Networks

Two artificial neural network models are developed to predict the effect of corrosion behavior of different series of aluminum alloys when exposed to two corrosive environments. Given the corrosion environment and time of exposure the first neural network predicts the ASTM G34 corrosion rating and the resulting material loss. The trained and limited test results predicted from this network are in good comparison to the experimental data. The effects of corrosion environment and material type from neural network simulation are presented to illustrate the trends. The second network predicts the cycles for final fatigue failure and the residual static strength of a particular type of material, given the amount of material loss due to corrosion. Based on the preliminary results, the neural network approach to corrosion and fatigue predictions is encouraging and can be used for a variety of materials and environments if more data is available. It is intended that the approach developed here will assist in the structural integrity evaluation of corrosion in aging aircraft.

7.34 Propagation from Corrosion to Structural Fatigue

The growth of general or surface corrosion is not well characterized in literature, and as such, the effects are modeled for a scenario as described using Fig. 7.48. The series of snapshots in time pictorially describes degradations and crack extensions

Fig. 7.48 Corrosion scenarios in lap splices

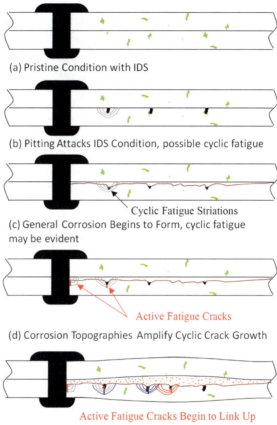

caused by time aging and cyclic loading. Figure 7.48a represents the section of a skin with an initial collection of material discontinuities scattered through the thickness, as received from the manufacturing floor. Figure 7.48b represents a point in time when the environment produces a chemical reaction at some of the near surface discontinuities, Initial Discontinuity State (IDS) sites, forming pits, which grow due to corrosion. Assuming low operational stresses, crack extension due to cyclic fatigue may or may not occur to any measurable amount. Figure 7.48c shows the general surface corrosion with slight material loss and a topography generated by accelerated discontinuity growths due to corrosion decay, and potentially, the early stages of cyclic fatigue. Figure 7.48d shows the topography roughness produces local micro stress amplifications that increase the effective crack stress intensities to result in crack extensions due to cyclic fatigue that otherwise could not be measured. The last lap in Fig. 7.48e illustrates the by-products produced by the corrosion begin to interject sustained stresses that can then produce sustained stress cracking extensions due to time and also aggravate the cyclic fatigue crack extensions [41].

The final combined result is an unanticipated cracking, multi-site scenario, which occurs significantly earlier in the service life of the structure than would have been determined without considering age degradation. The effects of corrosion are not well known, but typically have not been associated in the past with structural damage when corrosion alone acts. However, with the recognition that corrosion alone may account for the presence of small widespread or multi-site flaws, the chance that corrosion acting alone will not affect the structural integrity is slight. Therefore, we acknowledge that corrosion acts independently of fatigue, but also that corrosion and fatigue are interacting interdependently. These effects can be characterized by their impacts to the crack tip stress intensities. This allows introduction of families of β factors that capture not only finite width and depth effects but also local stress field changes such as topography, local effective geometric changes such as area loss, and induced sustained bending and tension stresses caused by time-dependent phenomena such as corrosion pillowing [42].

7.35 Direct Tension Stress-Corrosion Testing

Axially loaded tension specimens provide one of the most versatile methods of performing a stress-corrosion test because of the flexibility permitted in the choice of type and size of test specimen, stressing procedures, and range of stress levels [4]. The uniaxial stress system is simple; hence, this test method is often used for studies of stress-corrosion mechanisms. This type of test is amenable to the simultaneous exposure of unstressed specimens (no applied load) with stressed specimens and subsequent tension testing to distinguish between the effects of true stress corrosion and mechanical overload [43].

7.36 Considerations

There are several factors that may introduce bending moments on specimens, such as a longitudinal curvature, misalignment of threads on threaded-end round specimens, and the corners of sheet-type specimens. The significance of these factors is greater for specimens with smaller cross sections. Even though eccentricity in loading can be minimized to equal the same standards accepted for tension testing machines, inevitably, there is some variation in the tensile stress around the circumference of the test specimen which can be of such magnitude that it will introduce considerable error in the desired stress. Tests should be made on specimens with strain gages affixed to the specimen surface (around the circumference in 90° or 120° intervals) to verify strain and stress uniformity and determine if machining practices and stressing jigs are of adequate tolerance and quality.

Fig. 7.49 Spring-loaded stressing frame

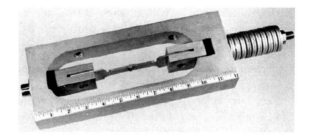

In actual testing with various types of stressing frames, such as the one shown in Fig. 7.49, the increase in net section stress will be somewhere in between. When the net section stress becomes greater than the nominal gross section stress and increases to the point of fracture, either of two events can occur: (1) fracture by mechanical overload of a material that is not susceptible to stress-corrosion cracking, or (2) stress corrosion cracking of a material at an unknown stress higher than the intended nominal test stress. The occurrence of either of these phenomena would interfere with a valid evaluation of materials with a relatively high resistance to stress corrosion. These considerations must be taken into account in experiments undertaken to determine "threshold" stresses.

7.37 Evaluation/Inspection

When stress-corrosion cracking occurs, it generally results in complete fracture of the specimen, which is easy to detect. However, when there is some uncertainty as to the presence of cracks due, for example, to the presence of corrosion products on the specimen surface, it may be necessary, at the conclusion of the test, to chemically clean the specimen to facilitate adequate inspection. It must be emphasized that fracture of the test specimen does not necessarily signify that stress-corrosion cracking has occurred. With specimens stressed by constant load, severe localized or generalized corrosion can lead to mechanical fracture by simple reduction of the cross-section area. While this can also happen with constant strain loaded specimens as a result of severe localized pitting corrosion, it is not likely to happen as a result of severe uniform corrosion. Also, constant-strain loaded specimens not having fractured may contain stress-corrosion cracks. Numerous small cracks developing in close proximity may cause relaxation of the stress. In such cases, metallographic examination can be used to establish whether or not there is stress-corrosion cracking present.

Tension tests of replicate specimens exposed with no applied stress, in conjunction with stressed specimens, can provide useful assistance in evaluating stress-corrosion effects, especially when stressed specimens do not fracture. In continuously increasing strain tests, the ultimate tensile strength, elongation, or reduction of area, or all three, should be measured. In addition, because complete fracture occurs with or without stress-corrosion cracking, a metallographic examination or other test should be performed to establish whether there is stress-corrosion cracking.

Performance and effectiveness metrics We define and use performance metrics for all major modules of the corrosion data analysis, detection and prediction architecture. Correlation metrics are defined for the optimum selection and extraction of CIs; confidence and false alarm rates are exploited to ascertain that fault/degradation detection results meet customer specifications; several metrics are defined for prognostic routines including Prognostic Horizon (PH) which identifies whether an algorithm predicts within a specified error margin; a metric called α-λ performance goes further to identify whether the algorithm performs within desired error margins; Relative Accuracy (RA) and Cumulative Relative Accuracy (CRA) that assume that prognostics performance improves as more information becomes available (Figs. 7.50, 7.51, 7.52, 7.53 and 7.54).

7.38 The Reasoning Paradigm: Dynamic Case Based Reasoning—The "Smart" Knowledge Base

A reasoning paradigm called Dynamic Case Based Reasoning (DCBR) that stores cases, matches new cases with stored ones and exhibits attributes of learning and adaptation will be used as the "smart" knowledge base to support learning and adaptation while providing the operator/maintainer the ability to interpret automated system outputs correctly and to effectively control the decision making

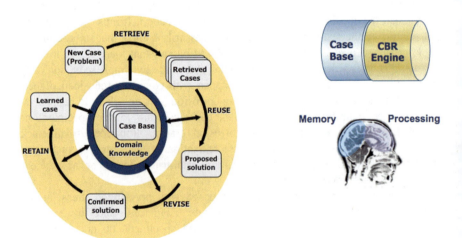

Fig. 7.50 The case based reasoning as a cognitive processing unit

7 Corrosion Diagnostic and Prognostic Technologies

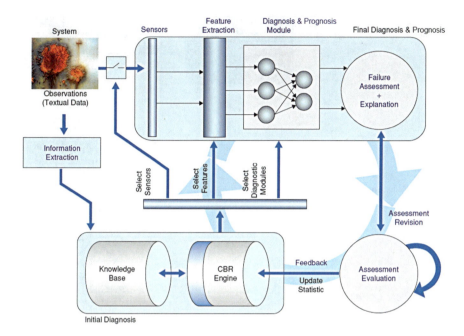

Fig. 7.51 The DCBR framework

process. We view this dynamic and generic knowledgebase and associated exploitation and control tools as an essential, novel, and effective way to link and exploit the human-machine information sources maximally while it serves as the "smart" strategy for accurate and robust degradation detection, prediction and fault-tolerance. We pioneered the development and implementation of DCBR in fault detection and isolation of critical aircraft components.

The DCBR interprets from sensor data not only static features but also dynamic and composite ones. The system module of the framework contains analytical models and algorithms that may be employed for dynamic and composite feature interpretation. Instead of one indexing path, the DCBR applies two—the abnormal symptom (AS) path, i.e. problem situation detected, and the problem description (PD) one. Furthermore, it entails functions to support case similarity evaluation and situational prediction through temporal reasoning and time tagged indexes. The remembrance calculation module updates the remembrance factors of existing cases.

The DCBR scheme offers significant advantages over conventional CBR systems such as:

- Combines the conventional case base techniques with the accumulated experience from the past maintain currency and efficiency.

Fig. 7.52 Case library format

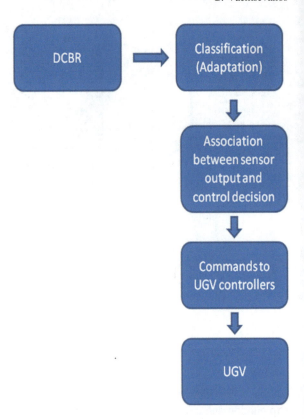

- Instead of learning from a problem situation over a long period of time, learning from multiple situations can be achieved in shorter time, serving several systems simultaneously.
- Suggests appropriate solution scheme, based on past successes, rather than just storing a new case.

Figure a typical case in the case library.
The cases in the case library according to following scheme:

- If ten predicted condition has an erre larger than 20%, it will be added to the case library.
- C = j, which satisfies $P_j > 0.5$
- Update cw for each case: $cw = \dfrac{\sum_{i \in G_j} Case_{i.cw}}{card(G_j)}$

Which represents the mean value of the weight in the set.

- If there is some case k such that $d_k = 0$. Then the measured features are exactly the same with one case form the library. Thus we can directly use the condition indicated by case k.

7 Corrosion Diagnostic and Prognostic Technologies

Case # j		
Case parameters	Alias	Description
E1	Casej.f₁	
O1	Casej.f₂	Anisotropy of the Energy and total Energy at each level of the wavelet decomposition
E2	Casej.f₃	
O2	Casej.f₄	
E_A	Casej.f₅	
Condition	Casej.C	Classify the image into baseline and pitted
Weight	cw	Evaluate how important is the case, initially set as 1

Fig. 7.53 A case in the case library

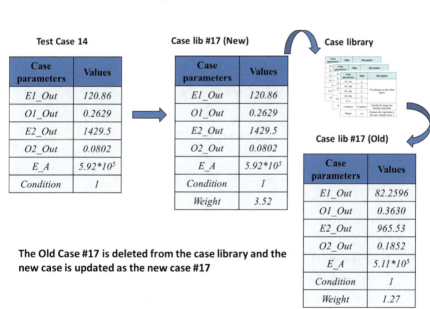

The Old Case #17 is deleted from the case library and the new case is updated as the new case #17

Fig. 7.54 Case update example

- Step 1: Evaluate the weight of different features
  - As some of the data are more related to the output and some others have a smaller correlation, we first do the correlation analysis and get the following weight:
  $$w_i = R(\vec{f_i}, \vec{C}) \quad for\ i = 1,...,5$$
  - Thus larger values are assigned to the parameters that are more likely related to the output.
- Step 2: Measure the distance between a new measurement of features f1,...,f5 denoted as measure.fi for i = 1.....5 and the features in each case in the Knowledge Base.
  $$d_i = \sum_{j \in S_1} w_j \cdot \left(\frac{measure.feature_j - case_i.feature_j}{measure.feature_j} \right)^2 + \sum_{j \in S_2} w_j \cdot case_i.feature_j^2$$

$S_1 = \{j | measure.feature_j \neq 0\}$
$S_2 = \{j | measure.feature_j = 0\}$
$S_1 \cup S_2 = \{1, 2, \cdots N\}$

Fig. 7.55 DCBR—distance metric for case adaptation

- If there is no case k such that $d_k = 0$ then determine the probablity of the measured features belonging to each condition:

$$P_i = \sum_{j \in G_i} cw_j \times p_j, \quad for\ i = 0, 1$$

$$G_i = \{k | case_k.condition = i\}$$

$$p_i = \frac{\left(\frac{1}{d_i}\right)}{\sum_{j=1}^{N} \left(\frac{1}{d_j}\right)}, \quad for\ i \in \{1, 2, \cdots N\}$$

7.39 Incremental Learning–The Reinforcement Learning Tool

Incremental learning will be pursued using Q-Learning, a popular reinforcement learning scheme for agents learning to behave in a game-like environment. Q-Learning is highly adaptive for on-line learning since it can easily incorporate new data as part of its stored database. An attractive feature in a game-like situation is that the player is learning to choose the best action for each particular game setting. In this framework, the expected reward or "cost-to-go" is stated as we also explore the utility of neuro-dynamic programming —another reinforcement learning tool—that employs a neural network to approximate the "cost-to-go" function; this technique addresses efficiently computational complexity issues. Incremental learning occurs whenever a new case is processed and its results are identified. Thus, the memory keeps track of each of its experiences, whether success or failure, in a declarative way; it is then ready to take advantage of future experiences. Figure shows a typical example of a case updating scheme.

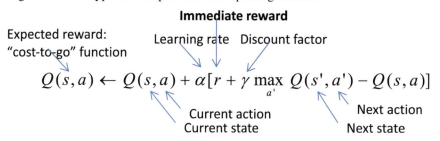

7.40 The Association Strategy—Relating Sensor Outputs (Features) to Control Decisions

Various algorithms are available to carry out efficiently and effectively the classification task. We will explore an "intelligent" method to map sensor outputs in the form of features to action decisions. The proposed approach will rely on a Dynamic Bayesian Network (DBN) as the association paradigm. Dynamic Bayesian Networks (DBNs) are developed for explicating the dynamic cause-effect relationships with the strength of the causal connection amongst the problem situations. DBNs are inference networks that use cyclic, directed graphs representing the causal relationships between concepts. DBNs can easily represent structured knowledge, especially for those containing causal relationships compared with neural networks. Each DBN represents the knowledge of an expert. Several DBNs can be combined into one by merging their adjacency matrices with different, weighted coefficients that usually represent beliefs of different human experts.

The key to stochastic modeling is to process data/information that is directly relevant to the goals at hand.

- Perception. This is the most primitive level of situational awareness. Providing more data of better accuracy in shortened times can improve level 1 situational awareness.
- Comprehension. In level 2 situation awareness, a decision maker integrates a variety of data with the goals and constraints they are operating under to develop an assessment of the situation.

- Projection. This highest level of situation awareness determines the best or most appropriate response by predicting the resulting courses of actions.

7.41 Performance Evaluation

We will establish and exercise appropriate performance metrics at each level of the proposed learning and decision support hierarchy.

The DCBR performance will be evaluated on the basis of the following criteria:

- Total solution search time—total time to search and rank matching cases.
- Quality of solution—relevance of retrieved cases; success and failure record; success of the top ranked case; overall success if several iterations are required.
- Comparison to legacy systems.

With regard to the incremental learning process, the emphasis, from a performance evaluation perspective, is on *correctness*. We will define appropriate correctness, precision and recall metrics for this purpose.

7.42 Performance Metrics/Specifications/Constraints

We will be defining generic performance metrics and specifications such as:

- Computational Speed—Data processing and image analysis algorithms must be executed in an expedient manner. It is crucial that performance metrics be defined and requirements specified on computational speed. Such metrics will dictate necessary tradeoffs between accuracy and algorithm sophistication and execution time.
- Accuracy—Accuracy metrics are defined relating to data mining, corrosion detection and prediction.
- Precision—The ability to pinpoint and resolve the existence of degrading conditions.

7.43 Epilogue

This chapter introduces a novel and comprehensive framework for corrosion health assessment, integrating robust corrosion testing and monitoring, data mining, corrosion detection, and prediction of corrosion damage growth, with intelligent reasoning paradigms. It is well documented that aircraft corrosion is a major concern that accounts for billions of dollars each year in efforts to detect, quantify, and prevent damage due to corrosion. Although significant advances have been reported over the recent past, there is still an urgent need for new technologies for sensing, data processing, and diagnostic/prognostic algorithm development aimed to provide crucial information to the aerospace and industrial process communities of impending structural failures and a means to mitigate them. The chapter addresses the introduction of new methods and tools essential for testing and data processing of corroding panels/parts; such tools become inputs to corrosion diagnostic and prognostic routines. A multitude of challenges remain to be addressed, major among them being the need for accurate and reliable sensing modalities, high fidelity corrosion models and integration architectures for better corrosion detection, prediction and prevention with an ultimate objective to reduce costs and improve the performance of aerospace and industrial assets.

References

1. López De La Cruz J, Lindelauf R, Koene L, Gutiérrez MA (2007) Stochastic approach to the spatial analysis of pitting corrosion and pit interaction. Electrochem Commun 9(2):325–330
2. Forsyth DS, Komorwoski JP (2000) The role of data fusion in NDE for aging aircraft. SPIE Aging Aircr Airpt Aerosp Hardware IV 3994:6
3. Brown DW, Connolly RJ, Laskowski B, Garvan M, Li H, Agarwala VS, Vachtsevanos G (2014) A novel linear polarization resistance corrosion sensing methodology for aircraft structure. In: Annual conference of the Prognostics and Health Management Society, vol 5, no 33
4. Brown D, Darr D, Morse J, Laskowski B (2012) Real-time corrosion monitoring of aircraft structures with prognostic applications. In: Annual conference of the prognostics and health management society, Mineapolis, MN, USA, Sept 23–27
5. McAdam G, Newman PJ, McKenzie I, Davis C, Hinton BR (2005) Fiber optic sensors for detection of corrosion within aircraft. Struct Health Monit 4:47–56
6. Vachtsevanos G, Lewis F, Roemer M, Hess A, Wu B (2006) Intelligent fault diagnosis and prognosis for engineering systems. Wiley, Hoboken
7. Orchard et al. (2005) A particle filtering framework for failure prognosis. In: World Tribology Congress III, Washington, D.C., Rep. WTC2005-64005
8. Kim H, Drake BL, Park H (2006) Adaptive nonlinear discriminant analysis by regularized minimum squared errors. IEEE Trans Knowl Data Eng 18(5)
9. Park C, Park H (2005) Nonlinear discriminant analysis using kernel functions and the generalized singular value decomposition. SIAM J Matrix Anal Appl 27-1
10. Wu B, Saxena A, Khawaja TS, Patrick R, Vachtsevanos G, Sparis P (2004) Data analysis and feature selection for fault diagnosis of helicopter planetary gears. IEEE Autotestcon
11. Wu B, Saxena A, Patrick R, Vachtsevanos G (2005) Vibration monitoring for fault diagnosis of helicopter planetary gears. IFAC Proc Vol (IFAC-Papersonline) 16:755–760
12. Orchard M (2007) A particle filtering-based framework for on-line fault diagnosis and failure prognosis. Ph.D. Thesis, Department of Electrical and Computer Engineering, Georgia Institute of Technology
13. Brown D, Abbas M, Ginart A, Ali I, Kalgren P, Vachtsevanos G (2010) Turn-off time as a precursor for gate bipolar transistor latch-up faults in electric motor drives. In: Annual conference of the Prognostics and Health Management Society, Portland, OR
14. Brown D, Edwards D, Georgoulas G, Zhang B, Vachtsevanos G (2008) Real-time fault detection and accommodation for COTS resolver position sensors. 1st international conference on Prognostics and Health Management (PHM), 6–9 Oct 2008

15. Straub D (2004) Generic approaches to risk based inspection planning for steel structures. Institute of Structural Engineering, Swiss Federal Institute of Technology, Zürich
16. Orchard M, Vachtsevanos G (2009) A particle filtering approach for on-line fault diagnosis and failure prognosis. Trans Inst Measurement Control 31(3–4):221–246
17. Orchard M, Vachtsevanos G, Goebel K (2011) Machine learning and knowledge discovery for engineering systems health management. In: Han J (ed) A combined model-based and data-driven prognostic approach for aircraft system life management. Chapman and Hall/CRC, Boca Raton, pp 363–394
18. Orchard M, Tang L, Goebel K, Vachtsevanos G (2009) A novel RSPF approach to prediction of high-risk, low-probability failure events. In: First annual conference of the Prognostics and Health Management Society—PHM09, San Diego, CA, USA
19. Engel S, Gilmartin B, Bongort K, Hess A (2000) Prognostics, the real issues involved with predicting life remaining. In: IEEE Aerospace. Big Sky, MT, pp 457–469
20. Lewis F (1986) Optimal estimation: with an introduction to stochastic control theory
21. Jardim-Gonçalves R, Martins-Barata M, Assis-Lopes J, Steiger-Garcao A (1996) Application of stochastic modelling to support predictive maintenance for industrial environments. In: IEEE international conference on systems, man, and cybernetics, pp 117–122
22. Groer P (2000) Analysis of time to failure with a Weibull mode. In: Maintenance and reliability conference, MARCON 2000
23. Frelicot C (1996) A fuzzy-based prognostic adaptive system. RAIRO-APII-JESA. J Eur Syst Autom 30(2–3):281–299
24. Ljung L (1999) System identification: theory for the user, 2nd edn. Prentice-Hall, New Jersey
25. Aha D (1997) Special issue on lazy learning. Artif Intell Rev 11(1–5):1–6
26. Studer L, Masulli F (1996) On the structure of a neuro-fuzzy system to forecast chaotic time series. In: International symposium on neuro-fuzzy systems, pp 103–110
27. Schwabacher M, Goebel K (2007) A survey of artificial intelligence for prognostics. NASA Ames Research Center
28. Li Y, Kurfess TR, Liang SY (2000) Stochastic prognostics for rolling element bearings. Mech Syst Signal Process 14:747–762
29. Muench D, Kacprzynski G, Liberson A, Sarlashkar A, Roemer M (2004) Model and sensor fusion for prognosis, example: Kalman filtering as applied to corrosion-fatigue and FE models. SIPS quarterly review presentation
30. Johnson S et al (2011) Prognostics. In: System health management with aerospace applications. Wiley, Chichester, Ch. 17, Sec. 1, pp 282–283
31. Pham H, Yang B (2010) Estimation and forecasting of machine health condition using ARMA/GARCH model. In: Mechanical systems and signal processing, pp 546–558
32. Tangirala R (1996) A nonlinear stochastic model of fatigue crack length for on-line damage sensing. In: Decision and control conference
33. Roemer M et al (2006) An overview of selected prognostic technologies with application to engine health management. In: Proceedings of GT2006 ASME Turbo Expo, Barcelona, Spain
34. Arulampalam S, Maskel S, Gordon N, Clapp T (2002) A tutorial on particle filters for on-line non-linear/non-Gaussian Bayesian tracking. IEEE Trans Signal Process 50(2):174–188
35. Orchard M, Tang L, Saha B, Goebel K, Vachtsevanos G (2010) Risk-sensitive particle-filtering-based prognosis framework for estimation of remaining useful life in energy storage devices. Stud Inf Control 19(3):209–218
36. Byington CS, Roemer MJ (2009) Selected prognostic methods with applications to integrated health management system. In: Applications of intelligent control to engineering systems, Springer, New York, Ch. 1, Sec. 1.6, pp 10–11
37. Schömig A, Rose O (2003) On the suitability of the Weibull distribution for the approximation of machine failures. In: Proceedings of the 2003 industrial engineering research conference, Portland, OR
38. Edwards D, Orchard M, Tang L, Goebel K, Vachtsevanos G (2010) Impact of input uncertainty on failure prognostic algorithms: extending the remaining useful life of nonlinear systems. In: Prognostics and health management conference

39. Saxena A, Celaya J, Balaban E, Goebel K, Saha B, Saha S, Schwabacher M (2008) Metrics for evaluating performance of prognostic techniques. In: Proceedings of international conference on Prognostics and Health Management
40. Saxena A, Celaya J, Saha B, Saha S, Goebel K (2009) On applying the prognostic performance metrics. In: Annual conference of the Prognostics and Health Management Society (PHM09), San Diego, CA
41. Craig HL, Sprowls DO, Piper DE (1971) In: Ailor WH (ed) Stress corrosion cracking, handbook on corrosion testing and evaluation. Wiley, New York, pp 231–290
42. Katwan MJ, Hodgkiess T, Arthur PD (1996) Electrochemical noise technique for the prediction of corrosion rate of steel in concrete materials and structures. Mater Struct 29 (5):286–294
43. Brooks C, Peeler D, Honeycutt KT, Prost-Domasky S (1998) Predictive modeling for corrosion management: modeling fundamentals. In: Predictive modeling for corrosion management: modeling fundamentals, Corfu Greece
44. Kumar S, Vichare N, Dolev E, Pecht M (2012) A health indicator method for degradation detection of electronic products. Microelectron Reliab 52(2):439–445
45. Lee S, Younsi K, Kliman G (2005) An online technique for monitoring the insulation condition of ac machine stator windings energy conversion. IEEE Trans Energy Convers 20 (4):737–745
46. Zhang B, Sconyers C, Byington C, Patrick R, Orchard M, Vachtsevanos G (2008) Anomaly detection: a robust approach to detection of unanticipated faults. In: IEEE conference on prognostics and health management, Denver, CO

Chapter 8
Assessing the Value of Corrosion Mitigation in Electronic Systems Using Cost-Based FMEA—Tin Whisker Mitigation

R. Bakhshi, P. Sandborn and E. Lillie

Abstract Pure-tin platings, which have become prevalent in electronics that uses lead-free solders, result in the spontaneous growth of conductive tin whiskers. If the tin whiskers bridge the gap between conductors, they can cause short circuit failures in systems. In this chapter we use cost-based FMEA to determine the projected cost of failure consequence for corrosion mitigation for the assembly of electronic systems. A case study of the lead-free implementation of a power supply demonstrates the return on investment of the control plan for the same product under various risk scenarios.

Keywords Electronic systems · Cost · Tin whiskers · FMEA—failure mode and effects analysis · Failure severity modelling

8.1 Introduction

Reliability is the ability of a system to perform its desired tasks within defined specifications for a specific period of time. A failure is the occurrence of an event that prevents the system from performing as it is intended to. Failures can be associated with a likelihood of occurrence and the consequences associated with them. Reducing the likelihood of the failure avoids the life-cycle costs of resolving those failures. Depending on the product and its support agreements, these are costs avoided by the system's owner, manufacturer or warranty service providers. However, reducing the likelihood of failure also costs money and a cost-benefit analysis is needed to investigate whether the investment in mitigation methods that improve the reliability have a financial benefit.

The goal of this chapter is to develop and analyze the cost trade-offs of implementing methods that mitigate tin whiskers, a corrosion-related failure

---

R. Bakhshi · P. Sandborn (✉) · E. Lillie
University of Maryland, College Park, USA
e-mail: sandborn@umd.edu

mechanism. The objective is to understand whether the mitigation methods will improve the reliability and subsequently lower the risk and if so, are the mitigation methods cost effective? Alternatively, if the adoption of the mitigation methods are mandated by the system's customer, will they improve the cost of ownership for the product throughout its useful life or not? We would also like to understand how the cost ramifications for the adoption of the mitigation methods change when the application of the product changes, in other words, how does the business case change when the risk environment changes.

Section 8.2 discusses corrosion mechanisms relevant to electronics focusing on tin whiskers and describes a model for whisker growth. In Sect. 8.3 we provide details of the risk model used in this chapter. Section 8.4 provides a case study of a printed circuit board used in a desktop PC and an avionics application for two different conformal coatings—Silicone and Parylene-C.

8.2 Tin Whiskers

There are several corrosion-related failure mechanisms that are relevant to electronic systems, these include surface contamination, electrochemical migration, and corrosion of metallization. The particular corrosion-related failure mechanism that we will focus on in this chapter is the growth of tin whiskers. Pure tin plating can result in spontaneous growth of conductive needle like structures that can be several millimeters long. Tin whiskers can create unintended bridges between conductive elements of electronics such as the leads on surface mount electronic components. They can also break and fall onto the printed circuit board surface and create an electrical connection between traces. The failures due to tin whiskers are hard to detect and sometimes result in very expensive failures in mission critical applications [1]. For example, a tin whisker can form and create a short circuit but the high current that passes through it vaporizes the whisker and the system can return to normal operation. These types of failures are intermittent failures, which are harder to detect than permanent failures [2]. Figure 8.1 shows tin whiskers that grew on leads of a surface mount component after exposure to a corrosive environment during an environmental test.

There is no conclusive explanation on why tin whiskers form. According to researchers in NASA [3], some theories believe that tin whiskers are formed due to stress relief in the component's tin plating while other theories believe that tin whiskers are due to changes in the grain structure of tin as a result of the plating process. Regardless of the root cause, tin whiskers pose risks to the reliable operation of electronic systems.

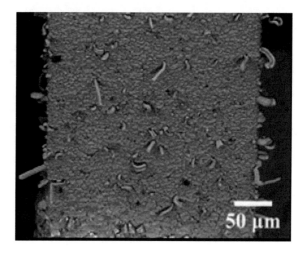

Fig. 8.1 Growth of tin whiskers under corrosive environment [11]. © 2012 Springer. Reprinted, with permission

8.2.1 Tin Whisker Mitigation

Historically, solder was composed of tin and lead, the addition of the lead stopped the growth of tin whiskers. As a result of environmental legislation in the early 2000s [4], lead was removed from solder (creating lead-free solders), and thus tin whiskers became a problem. There are several studies that have focused on tin whiskers, Han et al. [5] used breakdown voltage measurements to study the electrical short propensity caused by tin whiskers. Ashworth and Dunn [6] studied whisker growth phenomena in a series of experiments that lasted for 32 years. In another study, Han et al. [7] investigated the possibility of formation of metal vapor arcing due to tin whiskers. Their work showed that the electrical resistance of tin whiskers is a major factor in the formation of metal vapor arcs and proposed a metric to characterize the potential for vapor arc formation. Panashchenko and Osterman [8] investigated the use of a nickel under layer between copper and tin as a mitigation method to affect the whisker growth. Their efforts show that the environmental condition, mainly the corrosive environment has a larger impact on whisker growth than the nickel under layer. Mathew et al. [9] reviewed some of the mitigation methods to manage the growth of tin whiskers. They looked into conformal coating, plating techniques, the addition of under-layer material and annealing as four possible mitigation methods. In a later more comprehensive study, Zhang et al. [10] reviewed the concept of whisker growth, and discussed models that predict the growth mechanisms of whiskers. Their study covered several mitigation methods mentioned in the literature. These mitigation methods include:

- Avoiding pure tin plating: adding silver in the plating can reduce the propensity of tin whiskers.
- Solder dipping: although efficient, has side effects such as package cracking, loss of hermiticity and popcorning effects.

- Matte finish instead of bright tin: larger grain size and different carbon content of matte tin plating helps with reducing tin whisker growth.
- Extra under layer finish: having a thin layer of nickel underneath the tin plating decreases the compressive stresses that is believed to be the source of whisker growth by some researchers.
- Increase the thickness of the tin finish: there are mixed results as to whether thin or thick plating are better for reducing whisker growth. Overall, thin plating is not a common practice by industry due to its poor resistance to corrosion.
- Reducing compressive loads: some studies believe compressive loads are the reason behind the whisker growth, therefore reducing these loads by avoiding some loadings such as mechanical bending will mitigate the growth of whiskers.
- Heat treatment: processes such as reflowing, fusing and annealing tend to relieve stress and increase the grain size and subsequently reduce the growth of whiskers.
- Conformal coating: coatings can delay the growth of whiskers but will not eliminate them. Coatings can confine the majority of whiskers within the coating and prevent circuit shorts due to whisker growth.

Han et al. [11] conducted a series of experiments to investigate the growth of whiskers in corrosive environments for test specimens with several different conformal coatings (and without coatings as a benchmark). The results show that the corrosive environment will increase the whisker growth, i.e., both the density of whiskers and the average length of whiskers. The result of conformal coating was mixed. Some types of conformal coating showed whisker lengths longer than cases where no conformal coating was applied. However, two types of conformal coating showed promising results in terms of mitigating whiskers. Silicone conformal coating resulted in shorter whiskers. Parylene-C coating showed no whisker growth and completely eliminated the whisker formation. The results of this study will be the basis of our analysis on whisker growth in corrosive environments and we will consider Silicone and Parylene-C coatings as two effective mitigation methods to affect the growth of tin whiskers on electronic components.

8.2.2 Whisker Growth Modeling

Osterman et al. [12] provided a risk assessment methodology for tin whiskers by creating a stochastic tin whisker growth model. The risk associated with tin whiskers is the probability of a conductive whisker growing across the gap between two adjacent conductors that should be electrically isolated. The risk analysis developed in [12] is based on the growth characteristics of whiskers, the geometry of the product (the adjacent conductors) at risk, and time. The parameters that characterize whisker growth are the whisker density, whisker length, and the whisker growth angle. The whisker growth angle (Fig. 8.2) is the angle between a whisker and the surface from which the whisker grows [13]. All of the growth

Fig. 8.2 Tin whisker geometry definitions

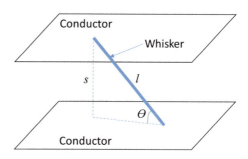

parameters are modeled as time-dependent probability distributions [14] that are determined from experimental data. The relevant product geometry includes the spacing between two adjacent conductors and available conductor surface area from which whiskers can initiate.

The spacing between two adjacent conductors is bridged (causing an electrical short circuit and thus a failure) if one whisker spans the gap (s) between the conductors. If l is the whisker length, the gap is spanned when,

$$l \sin(\theta) \geq s \qquad (8.1)$$

where θ is the whisker's growth angle. The criteria for bridging the gap between the adjacent conductors can, in general, be applied to any surface shape.

The algorithm (described in detail in [12]) for determining conductor bridging risk is to sample a whisker density distribution to determine the number of whiskers that will initiate in a given cross-sectional area (conductor surface area) and then sample each whisker's angle (Θ)—these quantities are assumed to be independent of time. For a particular time (duration), the whisker's length (l) can be sampled and used to determine if the whisker bridges the conductor spacing, i.e., the number that satisfy Eq. (8.1). If N is the sample size for Monte Carlo simulation, the risk of failure due to tin whiskers for part i, P_i, is defined as the ratio of the number of times the population of whiskers has at least one whisker that spans the gap (f) to the number of potential failure opportunities at a particular time. The final risk at a particular time is:

$$P_i(t) = \frac{f(t)}{N} \qquad (8.2)$$

If there are multiple instances of a particular part in a product, the risk for that part is estimated by

$$P^i_{Risk} = 1 - (1 - P_i)^{n_i} \qquad (8.3)$$

where i is the part, n_i is the number of instances of the part, P_i is the risk for one part; and P^i_{Risk} is the total risk for all the instances of the part. For m unique parts in a product, assuming no redundancy, the total risk from tin whiskers for the product is:

$$P_{Product} = 1 - \prod_{j=1}^{m}\left(1 - P_{Risk}^{j}\right) \tag{8.4}$$

8.3 Failure Severity Modelling

A model previously developed in [15] for the assessment of lead-free solder control plans has been adapted for the assessment of tin whisker mitigation approaches. This model is described in Sects. 8.3.2–8.3.5.

8.3.1 Cost of Reliability Models

Barringer [16] defines the cost of reliability as those costs that are used to keep the system free from failure. Models that estimate the cost of reliability based on Barringer's definition include [17, 18]. Models based on the risk of failure where failures are ranked based on severity and likelihood of occurrence have also been developed. Hauge and Johnston [19] define risk as "the product of the severity of a failure and the probability of that failure's occurrence". In [19], the severity and occurrence ratings are multiplied together to give a total magnitude of the risk due to the failure. Perera and Holsomback [20] describe a NASA risk management approach, which prioritizes risks based on likelihood and severity, with equal weight given to both factors. Perera and Holsomback identified risks from "fault tree analysis results, failure modes and effects analysis (FMEA) results, test data, expert opinion, brainstorming, hazard analysis, lessons learned from other project/ programs, technical analysis or trade studies and other resources". Sun et al. [21] describe a software cost of reliability model that incorporates the severity level of failures. They claim that the risk from a defect in software depends on both the failure rate of the defect and the severity level of the defect. According to Sun et al., the risk of a defect is defined as "the expected loss if [the defect] remains in the released software". Another concept introduced in the literature is the cost of risk. Liu and Boggs [22], in their paper on cable life, define the cost of risk as "the cost to a [electric] utility associated with early cable failure" and the cost of failure as "the cost to replace the cable". Liu and Boggs define the cost of risk as the cost of failures that occur before the end of the service life of the product.

Rhee and Ishii [23] introduced a cost-based failure modes and effects (FMEA) approach to measure the cost of risk and apply it to the selection of design alternatives. Kmenta and Ishii [24] use scenario-based FMEA to evaluate risk using probability and cost. Scenario-based FMEA uses predicted failure costs to make decisions about investments in reliability improvement versus maintenance. Taubel [25] implements a similar approach for the calculation of a total "mishap cost" by relating the known costs associated with mishaps to the probability of mishap for different severities of mishap. In Taubel's model, the definition of mishap derives from the Department of Defense's Military Standard 882C [26]: "an unplanned event or series of events resulting in death, injury, occupational illness, or damage to or loss of equipment or property, or damage to the environment".

The models developed in [23–25] form the basis for the model used in this chapter, which is described in the next section. We have extended these models so there is uncertainty in the life-cycle cost of the system and effectiveness of the technologies in reducing failures. The model in this chapter also replaces the FMEA probability of occurrence with discrete-event simulation based reliability sampling.

8.3.2 Failure Severity Model

In order to assess the cost of risk associated with whisker growth mitigation, we will determine the difference in failure consequence costs between the system with and without the mitigation. Note that the method described in this section does not calculate the actual life-cycle cost of the system, but rather the cost difference between the resolution and consequences of failure for the two cases while assuming that all other life-cycle cost contributions are a "wash". This is referred to as a "relative accuracy" cost model in [27].

Systems can fail in different ways, and all failures do not necessarily have the same financial consequences. A system failure that requires maintenance (repair) might cost less than a failure that requires the system owner to replace the system. Ideally the system owner needs to predict the cost of all the failure events that are expected to occur over the life of the population of systems, considering that those systems can fail multiple times, in multiple ways, and with different financial consequences of failure depending how the systems fail.

Taubel [25] calculates a total mishap cost by plotting the known costs associated with mishaps versus the probability of mishap for different severities of mishap (e.g., Figure 8.3). In the model, each severity level has a distinct cost and an associated probability of occurrence. The area under the curve is the expected total mishap cost.

A mitigation activity is a process that may reduce the overall expected number of mishaps at specific severity levels. Each mitigation activity is assumed to affect a specified set of severity levels and does not change the probability of a failure for the other severity levels.

The model described in this section determines the expected number of failures at each severity level rather than calculating the probability of failure at each severity level. This is done because some failures may occur more than once during the life of the product, hence the cost of (multiple) failures is accounted for.

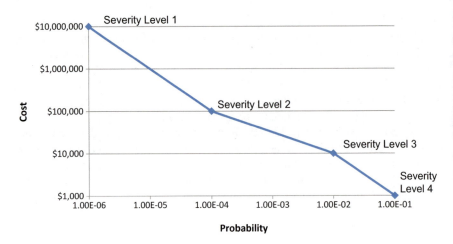

Fig. 8.3 Multiple severity model (after [25])

Fig. 8.4 Modeling steps [15]. © 2015 Elsevier. Reprinted, with permission

1. Determine all relevant failure modes

2. Determine the expected number of occurrences and costs per occurrence for each failure mode

3. Determine the total cost of failure

4. Select a set of mitigating activities

5. Determine the cost of performing the mitigating activates and repeat steps 2 and 3 with the mitigation activities used

6. Determine the ROI of performing the activities

We refer to this as the Projected Cost of Failure Consequences (PCFC) for the population (e.g., fleet) of products.[1] An overview of the steps in the model is shown in Fig. 8.4.

---

[1] To clarify, the models used in [25] and in this chapter (although not exactly the same—see Sect. 8.3.3) are continuous risk models, i.e., they assume that probabilities are continuous, therefore the PCFC is defined as the area under the curve. However, some risk models assume the probabilities are discrete, in which case the cost of failure would be calculated as the sum of the probability of failure at each discrete severity level multiplied by the cost of failure resolution at the corresponding severity level. Both approaches are valid, continuous risk is assumed in this chapter.

8 Assessing the Value of Corrosion Mitigation ...

The first step in the model is to identify and describe each relevant failure by determining the part affected by the failure, and the failure mode, cause, and mechanism associated with an occurrence of that failure. Additionally, each failure is defined by an application-specific severity level. The severity level determines the cost associated with an occurrence of the failure.

Next, the number of failures expected to occur over the service life of the product at each severity level are determined. The collective expected number of failures for each severity level is called the severity level profile. The calculation of the expected number of failures per product per unit lifetime for each distinct severity level is given by:

$$f_i = \sum_{j=1}^{n} f_j \qquad (8.5)$$

where f_i is the expected number of failures of severity i per product per unit lifetime; and n is the number of ways a product can experience failure at severity level i.

8.3.3 Determining the Initial PCFC

Assuming a repairable system, each failure experienced by the system is described by two characteristics: the severity of failure and the frequency of occurrence of that failure. Severity correlates to the cost of the actions that the system or product owner will have to take to correct or compensate for the effects of a failure after it has occurred. One possible source of data for determining a PCFC is a Failure Modes, Mechanisms, and Effects Analysis (FMMEA) report (e.g. [28]).[2]

Most FMMEAs qualitatively describe severity and frequency of failure, whereas to be used in this model each failure's severity and frequency must be quantitatively defined. Each failure's severity and frequency will be used to determine: (1) the expected cost that the system owner will incur for every instance of the occurrence of that failure, and (2) the number of times the failure is expected to occur over the service life of the system.

For example, in the FMMEA used for the case study in this chapter, severity of failure is rated on a scale of 1–5, with a severity 5 failure defined as a minor

---

[2]A FMMEA categorizes failure events and assigns each event a rating for its severity and likelihood of occurrence. Alternatively, a Failure Modes and Effects Analysis (FMEA) or a Failure Modes Effects and Criticality Analysis (FMECA) could also be used as a source of data for the severities and frequencies of the ways a system could fail. A FMEA is very similar to a FMMEA, except that a FMEA does not analyze the mechanisms associated with each failure. Additionally, a FMECA is an extension of a FMMEA that includes a criticality analysis. Criticality analysis is a method of prioritizing failures after each failure is assigned a severity and occurrence rating, where the highest priority failures (those to be dealt with first), are those with the highest aggregate severity and occurrence ratings.

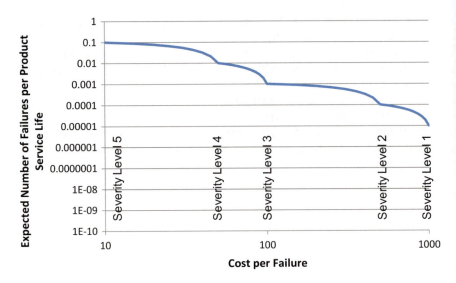

Fig. 8.5 Expected number of failures versus cost per failure [15]. © 2015 Elsevier. Reprinted, with permission

nuisance and a severity 1 failure defined as a catastrophic failure. Each of these severities must be assigned an expected cost associated with the consequences of the occurrence of a failure of that severity.[3]

The transformation of FMMEA ratings to numerical values of cost and expected number of failures is application specific. The cost associated with a specific severity of failure and expected number of failures for a given frequency rating could vary based on several factors including: operating conditions, the context the system is being used in, and the length of the service life.

Using an expected number of occurrences for each failure severity, and a cost associated with each occurrence of every failure, the PCFC for the system can be determined. Figure 8.5 shows a plot of the expected number of failures and cost associated with each failure for five severity levels. The vertical axis is the number of failures expected to occur per product per service life. The service life is the required life the system, expressed in years or temperature cycles. The horizontal axis is the cost per failure event.[4]

---

[3]It should be noted that FMMEAs also describe the frequency of failure on a qualitative scale (this is usually called the "probability of occurrence"). Kmenta and Ishii [24] use the probability of occurrence; however, in the model presented in this chapter, the expected number of failures per product per service life are determined from reliability distributions, not generated from the FMMEA.

[4]The model described in this chapter assumes that the cost of failure decreases linearly between severity levels. The assumed linear decrease appears as shown in Figs. 8.5 and 8.6 when graphed on a log-log plot. For the plots in the case study, the lines between severity levels are represented by straight lines (on the log-log plots) for graphical convenience.

The cost and number of failures for each severity level are connected and form a curve as shown in Fig. 8.5. The area under this curve is the PCFC for the system.

$$PCFC_{initial} = \int_{E_1}^{E_u} C(x)dx \qquad (8.6)$$

where E_1 is the expected number of severity level 1 failures (E_u is the expected number of severity level m failures); u is the number of severity levels under consideration and $C(x)$ is the cost of a failure event occurring at severity level x.

In practice the area of the discrete trapezoids formed by the points in the curve are determined and summed using,

$$PCFC_{initial} = \sum_{i=1}^{u} [E(i+1) + 0.5E(i)][C(i+1) - C(i)] \qquad (8.7)$$

where $E(x)$ is the expected number of failures per product per unit lifetime of point (severity level) x on the curve.

8.3.4 Activities Affecting the Number of Failures

An activity is sub-process, process, or group of processes that when performed (or applied) changes the expected number of failures over the service life of the product. Activities can be performed at multiple levels of rigor; rigor is the detail or depth at which the activity is performed. Performing an activity at a higher level of rigor has the potential for a greater reduction in the number of expected failures, but it will cost more.

Activities can affect specific failure modes, failure mechanisms, failure causes, and parts. If an activity affects the mode, mechanism, cause, or part that corresponded to a failure in the FMMEA used to create the initial severity level profile, then if that activity is performed, the expected number of failures will change. Equation (8.8) shows the calculation of the new expected number of failures after activities are performed.

$$N_{f-f} = N_{f-i} \prod_{i=1}^{q} P_R(i, R) \qquad (8.8)$$

where N_{f-f} is the number of failures expected to occur over the service life of the product for a particular failure listed in the FMMEA after considering activities; N_{f-i} is the number of failures expected to occur over the service life of the product for a particular failure listed in a the FMMEA before considering activities; $P_R (i, R)$ is the fractional reduction in the expected number of failures occurring over the service life of the product due to performing activity i; q is the number of activities performed that affect the failure under consideration; and R is the level of rigor activity i is performed at.

An activity is defined by the change in failures over the service life of the product, the non-recurring (NRE) cost for each level of rigor, and the particular failure modes, failure mechanisms, failure causes, and parts the activity will impact if performed.

The cost of performing all activities, called the Total Implementation Cost, (C_{Total}) is calculated according to,

$$C_{Total} = \sum_{i=1}^{q} C_{NRE}(i, R) \qquad (8.9)$$

where $C_{NRE}(i,R)$ is the cost of performing activity i at level of rigor R.

Performing an activity at level of rigor R may reduce the number of times a failure is expected to occur. The model determines which failures listed in the FMMEA each activity affects by checking if a failure's mode, mechanism, cause, and part are impacted by the activity. The model performs the calculation for each activity on every failure listed in the FMMEA whose mode, cause, mechanism, and part are all impacted by the activity.

Once a set of activities has been chosen, the model calculates the modified PCFC for the system. First the model calculates the number of failures expected to occur at each severity level using Eq. (8.5) and generates a modified severity level profile. Next, the model uses the new expected number of failures (determined via a discrete-event simulation that samples cycles to failure distributions through the support life of the product—see the case study) to calculate expected PCFC of the system using,

$$PCFC_{modified} = \int_{E_{1-f}}^{E_{m-f}} C(x)dx \qquad (8.10)$$

where E_{1-f} is the expected number of severity level 1 failures after activities are considered and E_{m-f} is the expected number of severity level m failures after activities are considered.

The difference between the initial PCFC and the modified PCFC, called the *Reduction in Failure Cost* is calculated as,

$$Reduction\ in\ Failure\ Cost = PCFC_{Initial} - PCFC_{Modified} \qquad (8.11)$$

The *Reduction in Failure Cost* can be graphically represented as the difference in the areas under the curves in Fig. 8.6. The top curve is the expected number of failures versus PCFC before activities are considered, and the bottom curve is the expected numbers of failures versus PCFC after activities are considered.

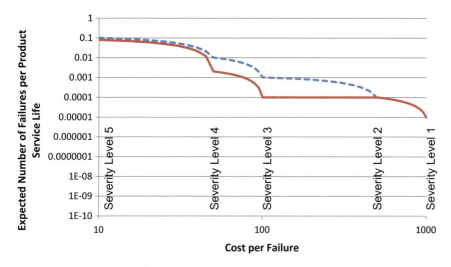

Fig. 8.6 The blue (dashed, top) curve represents the number of failures per product per unit lifetime at each severity level before activities are considered, and the red (solid, bottom) line represents the expected number of failures with the activities performed [15]. © 2015 Elsevier. Reprinted, with permission

8.3.5 Return on Investment

The final step in the model is to calculate the *Return on Investment* or ROI. The ROI is defined as the difference between return and investment divided by investment. In this model, the investment is the money spent on performing activities, the *Total Implementation Cost*, and the return is the PCFC that will be avoided because activities have been performed, the *Reduction in Failure Cost*.

$$\text{Return on Investment (ROI)} = \frac{\text{Reduction in Failure Cost} - C_{Total}}{C_{Total}} \quad (8.12)$$

8.4 Case Study—The Cost Implications of Implementing Whisker Growth Mitigation Plans

In this section, we will use the model described in Sect. 8.3 to calculate the cost implications of using mitigation approaches that will affect the growth of whiskers in electronic systems. We will consider a generic printed circuit board (PCB) with a congested layout of surface mount components. The case study will analyze two

Table 8.1 Surface mount components on the PCB

Number of leads on one side (number of sides)	Number of surface mount components on the PCB	Gap size (mm)	Severity level
38(4)	2	0.08	5
6(2)	10	0.12	1
12(4)	8	0.1	3
10(2)	8	0.12	2
19(4)	4	0.1	4

different applications of the PCB each with its own specific working environment conditions and risk consequences. The first application is a desktop computer with the expected life of 5 years while the second application is using the PCB in a commercial aircraft with the expected life of 20 years.

The PCB has different types of surface mount components soldered onto it. The leads on the surface mount components are the elements most susceptible to growth of whiskers. As mentioned in Sect. 8.2.2, the gap size between the leads is a major factor in formation of short circuit connection between two adjacent leads. Here, we only include the components with gaps between the leads that are narrow enough to form whiskers that can make a conductive bridge to an adjacent lead. The number of leads on each of these surface mount components varies. Some have leads only on two sides while some others have leads on all four sides. Table 8.1 summarizes the number of surface mount connections of various types on the PCB with gap size between adjacent leads and their associated severity level for a failure occurrence. A lower severity level number means failure is more serious and thus more expensive to resolve if it occurs.

It's important to note that since the whisker mitigation activities are only applicable to whisker growth, we won't be considering all the possible failures that may occur in the PCB, i.e., all the failures in an FMMEA analysis. We assume that the whisker mitigation plans will not affect the failures caused by the other failure mechanisms, i.e., with and without the tin whisker mitigation plans, the number of failures due to non-tin whisker causes remain the same for the application. We also assume that the type of failures that occur due to whiskers could be either permanent (hard failures) or intermittent (soft failures). As mentioned earlier, the high current in a short circuit caused by a whisker can vaporize the whisker and subsequently fully or partially restore the operation of the affected components in some cases. Regardless of whether the failure is permanent or intermittent, since the loss of functionality occurred, we count the failure and assume that the mitigation plan will affect the further occurrence of the failures. However, we are assuming that intermittent failures are less severe than permanent failures, hence they cost less.

8.4.1 Whisker Mitigation Activities—Conformal Coating

There are several approaches discussed in the literature to mitigate the whisker growth. In this work, we focus on one particular approach, which is conformal coating. Conformal coating is deposition of a layer of a chemical compound on a PCB and its components. The goal of conformal coating is to create an insulation layer that prevents exposure of electronic circuits to corrosive environments and moisture.

There are several methods to create a coating including: dipping, brushing and spraying (either by hand or by robots). The method depends on the type of coating, which itself is a function of the application of the boards. Different coating types provide different levels of insulation, which depend on the environmental conditions associated with the board's application. An ideal conformal coating, covers the sides and the edges thoroughly. In this chapter, we will focus only on Silicone and Parylene-C coatings since they are the two methods that showed the most promising results for whisker mitigation in the Han et al. [11] study.

8.4.1.1 Silicone Coating

Silicone coating can be applied to the electronic circuitry through spraying (hand or robot), dipping and brushing depending on the application of the boards and the level of masking. It provides a thin transparent layer with good resistance to moisture, high temperatures and airborne particles. For boards that require further work after the coating, Silicone is an ideal candidate because it has a weak resistance to solvents.

8.4.1.2 Parylene-C Coating

Parylene-C is considered the best conformal coating as it lacks many of the deficiencies of other coatings. Parylene-C forms at room temperature and therefore does not induce any stress on the components that it coats. It penetrates narrow gaps well and provides a covering for all the surfaces. It also provides a uniform thickness on all surfaces that is sometimes hard to achieve with other conformal coating materials.

However, the process of applying Parylene-C is more complicated than other types of conformal coating. Prior to coating, electronic components and connectors have to be carefully masked and the surface of the board has to be cleaned carefully and thoroughly. The coating process is performed in a chamber with specialized vapor-phase deposition techniques. This creates challenges since vapor can get into areas that should not be coated, hence the need for masking. The deposition of the coating on the surface takes place at a constant rate, so depending on the application of the board (i.e., the thickness of the coating required), the process can take from several hours to more than a day for a single batch of boards.

Parylene-C coating is very labor intensive since masking the components can take a lot of time. Programming a robot that will do the deposition is expensive as well. Overall, the Parylene-C coating may be an expensive option depending on the complexity of the boards, their application and the number of the boards that need to be coated.

8.4.2 Board Applications

In this section, we investigate two separate applications of the board. The first application is using the board in a desktop computer and the second application is using the board in systems used in a commercial aircraft. The layout of the board does not change, however, the operational environment conditions, length of operation, risks associated with failures and costs of failures are different for the two applications. A desktop computer is in an environment that is assumed to have stable temperatures, pressure, and humidity, while in an aircraft we are assuming that the board is operating in the unpressurized, non-climate controlled, tail of the aircraft, colloquially known as the "hell hole." The conditions in the hell hole are assumed to be those defined by [29].

For this case study, we assume that a typical commercial aircraft has an expected service life of 20 years. Alternatively, we assume that the desktop computer has an expected service life of 5 years. When used in an aircraft, we assume the board will experience one temperature cycle per flight, and that the aircraft is making an average of 6 flights per day, and that it operates 300 days per year. We assume that the desktop computer will encounter 1800 temperature cycles per year (on/off and sleep cycles). Table 8.2 summarizes the operational expectations of the board when used in both applications.

The consequences of a failure in an aircraft can be far greater than a failure in a desktop computer. In the context of this work, we consider the consequences of failure in terms of the financial loss to the entity or entities responsible for the performance of the system. For the desktop computer case, we assume the entity responsible for failure costs is the manufacturer, and the computer is under warranty for the service life (five years). In the commercial aircraft case, we assume the entity

Table 8.2 Usage conditions for the board

	PC	Commercial aircraft
Temperature cycles (per lifetime) (c)	9000	36,000
Service life (years)	5	20
Temperature cycles per year	1800	1800
Number of units in service	100,000	500

8 Assessing the Value of Corrosion Mitigation ...

Table 8.3 Consequences and likelihoods of varying severity and occurrence ratings of failure for the board in a desktop PC and a commercial aircraft

Severity of failure	Desktop computer		Commercial aircraft	
	Failure event associated with	Failure cost	Failure event associated with	Failure cost
5	Minor nuisance	$10	Minor nuisance	$100
4	Minor repair	$50	Minor repair	$2500
3	Board replacement	$150	Board replacement	$5000
2	Board replacement and collateral damage to PC	$400	Repair or replacement, interrupting flight schedule	$25,000
1	Loss of entire PC	$1200	Repair or replacement, causes collateral damage	$250,000

responsible for failure costs is the aircraft owner (the system operator). Table 8.3 shows the assumed consequences and likelihoods of varying severity and occurrence ratings of failure when the board used in a desktop computer and a commercial aircraft.

In this case study we assume that all repair and replace maintenance actions associated with a part result in a good-as-new part in the product. Also, the "minor nuisance" failure event could be a "no fault found" failure event. This case study assumes that the financial consequences of a no fault found event (severity level 5) do not change if multiple no fault found events occur on the same board.

In this case study, it is assumed that for tin whisker induced failures on surface mount components, the reliability can be modeled with a Weibull distribution:

$$F(c) = 1 - e^{-\left(\frac{c-\gamma}{\eta}\right)^{\beta}} \quad (8.13)$$

In Eq. (8.13), $F(c)$ is the cumulative distribution failure, c is the number of temperature cycles (Table 8.2); β is the Weibull shape parameter, η is the Weibull scale parameter, and γ is the Weibull location parameter. In this study, the location parameter is assumed to be zero. Table 8.4 shows the shape and scale parameters used in this study for the surface mount components on the board. The probability of a failure occurrence due to tin whisker formation is assumed to be a function of

Table 8.4 Weibull parameters for the components on the board

Number of leads on one side (number of sides)	Shape parameters	Scale parameters (cycles)
38(4)	2.9	50,000
6(2)	2.9	60,000
12(4)	2.9	55,000
10(2)	2.9	58,000
19(4)	2.9	52,000

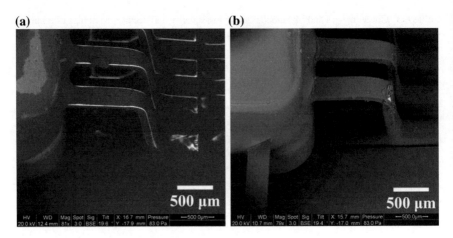

Fig. 8.7 Conformal coating on surface mount components. **a** Silicone, **b** Parylene-C [11]. © 2012 Springer. Reprinted, with permission

the board design (number leads on the surface mount, gap size between the leads, length and angle of whiskers) and therefore same reliability values are used for both applications. These values are derived from the model explained in Sect. 8.2.2.

A discrete-event simulator that samples the respective cycles to failure distributions for each of the product's parts was used to determine the sequence of failure events. The discrete-event simulator was run through the entire service life of the product to determine the total failure counts for each part in the product. A Monte Carlo model was used and 1000 independent time histories of the products analyzed in the case study were run to build the results provided.

As mentioned earlier, the control plans to mitigate the growth of tin whiskers are using two different conformal coatings on the boards, Silicone and Parylene-C coating. These coatings provide different levels of mitigation as outlined by Han et al. [11]. For the desktop computer, we only consider the Silicone coating while for the aircraft, we consider both Silicone and Parylene-C coatings. Figure 8.7 shows examples of surface mount components with Silicone and Parylene-C coatings.

The costs of the mitigation plan depend on the board's application, number of boards, and the contractor that performs the conformal coating. Therefore, we use probability distributions to calculate the costs of coating for the boards. Silicone coating is the less expensive option and it can cost as little as $2 per board. Parylene-C is more expensive due to the complicated process of coating, labor and programming of the coating equipment; therefore, the costs of coating have a wide range. Table 8.5 shows the distributions and their parameters that are used to calculate the costs of coating for Silicone and Parylene-C. These cost values were obtained through communication with companies that provide conformal coating service. Cost values represent the costs of performing the mitigation for the entire

8 Assessing the Value of Corrosion Mitigation ...

Table 8.5 Triangular distribution parameters used for cost and failure reduction calculations

	Conformal coating		Mode	Low	High
Costs of coating ($)	Silicone		400,000	200,000	500,000
	Parylene-C		375,000	250,000	500,000
Failure reduction coefficient	Silicone		0.8	0.7	0.9
	Parylene-C	Average	0.5	0.35	0.6
		Range	0.2–0.7	0.05–0.55	0.3–0.8

population of boards (i.e., 100,000 for desktop computer and 500 for commercial aircraft).

For the reliability improvements, probability distributions are used to quantify the effects of conformal coating on mitigating tin whisker failures. Table 8.5 also shows the parameters used to generate the reduction coefficients that will be used to calculate the failure reduction due to conformal coatings of the boards. These coefficients are multiplied by the number of failures that occurred in a board without conformal coating in order to calculate the number of failures when the board has conformal coating. Experiments by Han et al. [11] showed Parylene-C completely prevented the growth of tin whiskers during the experiments period. However, this may not be the case for the 20 year expected life of an aircraft. Therefore, we assume that there will be tin whisker growth when Parylene-C is used but the growth is limited. We consider a range of values for failure reduction coefficient in the case of Parylene-C coating and investigate the sensitivity of ROI values to changes in these coefficients.

Each application of the board was run for 1000 trials (life histories). Each trial calculates the initial PCFC by sampling the cycles to failure distributions for each component in the product. Each surface mount component must complete the number of cycles defined by the service life. If a component does not survive its service life a corrective action is taken (repair or replace) and the component samples the cycles to failure distribution again until the cumulative lives (in cycles) of each component is greater than or equal to the service life (also in cycles). For each trial we calculate: an investment cost (the cost of performing activities) by sampling the cost distribution defining the cost of performing the activity, a return (the reduction in PCFC after performing activities) by sampling the distribution defining the fractional reduction in failures for each activity performed and applying the factional reduction in failures to the failures in the FMMEA[5] that the activity affects, and an ROI. Thus, for each trial, the initial PCFC, investment cost, and return could be different because the parameters that determine them are defined as distributions that are sampled for each trial.

---

[5] As stated previously, we are only considered tin whisker failures in this chapter. All other failure mechanisms are assumed to be unaffected by the tin whisker mitigation and therefore are a "wash".

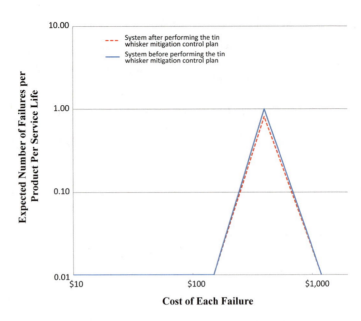

Fig. 8.8 Results of the model for a single Monte Carlo run on the desktop computer population with Silicone coating

8.4.3 Desktop Computer Application Results

In this section, the mitigation plan is applied to the board used in a desktop computer. The results of the case study are shown in Fig. 8.8, where the blue (solid) lines represent the system before the tin whisker mitigation control plan activities are performed, and the red (dash) lines represent the system after the tin whisker mitigation control plan activities are performed. Note that while 1000 trials were performed, for illustrative purposes, Fig. 8.8 only shows the results of a single trial. In this particular example, throughout the five-year lifetime of the desktop computers in the population, each computer has an average of a single failure occurrence due to tin whisker growth. This failure is a severity 2 failure where the cost of failure is $400.

For the 1000 trials of the model, very few had tin whisker related failures. Out of 1000 trials, only 132 of them showed any type of failure and the rest did not have any tin whisker related failures (Fig. 8.8 is one of the 132 trials that had tin whisker related failures). In cases where there were failures, on average no more than one failure was observed during the 5 years life cycle of a desktop computer (similar to the case in Fig. 8.8 where only a single failure was observed). In cases where no failure occurred, the ROI associated with the mitigation approach is −1 (i.e., the investment in tin whisker mitigation was made, but there was no return on the investment). Figure 8.9 shows the range of ROI values for this mitigation plan.

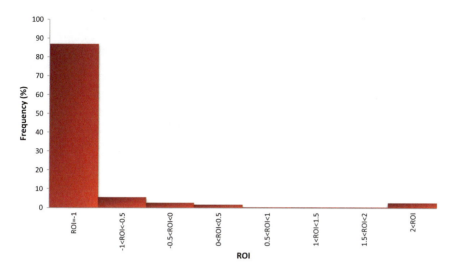

Fig. 8.9 Distribution of ROI values for Silicone coating in desktop computer

As the results show, in the case of the desktop computer application only 5% of cases yield a positive return for tin whisker mitigation using Silicone. The mean ROI for Silicone conformal coating (to mitigate tin whiskers) for the PC application was −78% indicating that from a financial point of view, this is obviously not worth doing.

8.4.4 Commercial Aircraft Application Results

8.4.4.1 Silicone Conformal Coating

For the commercial aircraft application, two cases will be considered. In one case the boards will be coated with Silicone and then for the second case, the effects of Parylene-C conformal coating will be examined. Figure 8.10 shows a single run of Monte Carlo over 20 years of operation. This is the expected number of failures per board for the whole population of aircraft.

For 20 years expected life of an aircraft, all 1000 trials showed tin whisker related failures. During this time, there were on average 4 failures due to tin whisker growth in an aircraft, which the Silicone coating managed to reduce. Figure 8.11 shows the average number of failures for the aircraft population in 5 Monte Carlo runs.

Figure 8.12 shows the distribution of ROI values for the Silicone coating case. In order to make the figure easier to read, all the ROI values are divided by 10^3, because of the very large values of ROI. This is because the costs of coating are insignificant relative to the costs of failures and their subsequent repairs in a commercial aircraft.

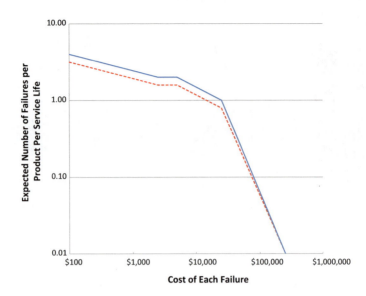

Fig. 8.10 Results of the model for a single Monte Carlo run on the aircraft population with Silicone coating

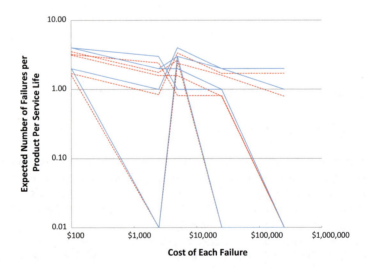

Fig. 8.11 Results of the model for a 5 Monte Carlo run on the aircraft population with Silicone coating

8.4.4.2 Parylene-C Conformal Coating

Although Silicone coating produces a large return on investment for mitigating tin whiskers, there may still be too many failures occurring due to tin whiskers for a

8 Assessing the Value of Corrosion Mitigation …

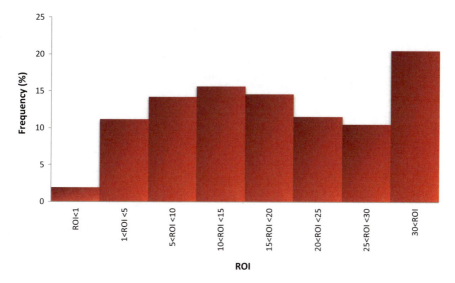

Fig. 8.12 Distribution of ROI values for Silicone coating in commercial aircraft. Values divided by 10^3

mission critical system such as a commercial aircraft. Parylene-C, which showed promising results in experiments becomes a viable option for coating the boards that are used in the commercial aircraft. Figures 8.13 and 8.14 show a single and 5

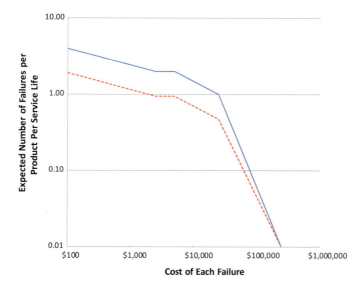

Fig. 8.13 Results of the model for a single Monte Carlo run on the aircraft population with Parylene coating

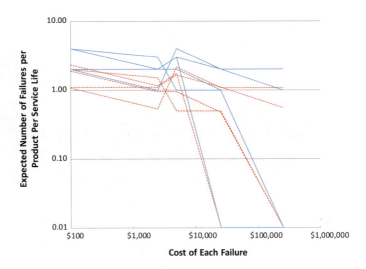

Fig. 8.14 Results of the model for a 5 Monte Carlo run on the aircraft population with Parylene-C coating

Monte Carlo runs respectively for the whole population of aircrafts using the average values in Table 8.5. The larger gap between the blue solid line and the red dashed line show that reliability has significantly improved by using Parylene-C.

The Parylene-C coating, though more expensive than Silicone, results in positive ROI values while at the same time significantly reducing the chances of failure due tin whisker failures. Figure 8.15 shows the distribution of ROI values for Parylene-C coating using the average values in Table 8.5.

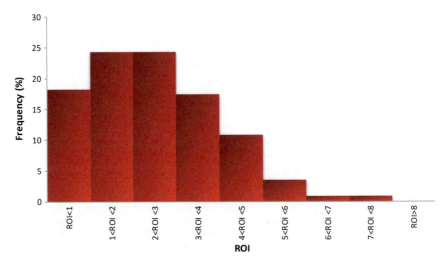

Fig. 8.15 Distribution of ROI values for Parylene-C coating in commercial aircraft. Values divided by 100

8 Assessing the Value of Corrosion Mitigation … 337

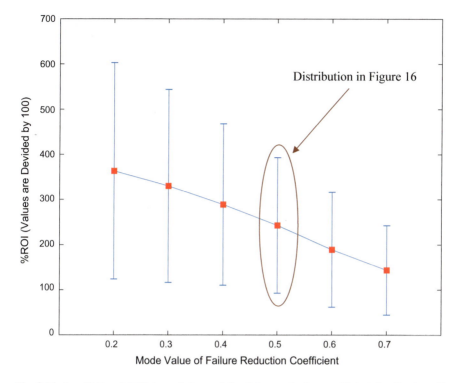

Fig. 8.16 Sensitivity of ROI to variations of the failure reduction coefficient for Parylene C coating

Figure 8.16 shows the sensitivity of the ROI values to a range of failure reduction coefficient values. These are the ranges in Table 8.5. As it can be seen a 50% improvement in reliability, yields an average ROI of 280. An 80% improved reliability (which corresponds to a 0.2 failure reduction coefficient) has the largest ROI values as it is expected.

8.5 Epilogue

Adoption and insertion of new technologies and processes into systems is inherently risky—in our case lead-free solders that create a tin whisker risk. An assessment of the cost of risk may be a necessary part of planning or building a business case to change a system. A cost-based FMEA model that forecasts the cost of risk associated with inserting a new technology into a system has been used to assess a tin whisker mitigation control plan for the same product in two different risk scenarios. In the model, the projected cost of failure consequences (PCFC) is defined as the cost of all failure events (of varying severity) that are expected to occur over the service life of the system. The PCFC is uncertain, and the potential positive impact of adopting new technologies into the system is to reduce the cost of risk and/or reduce its uncertainty.

The case study presented assesses the adoption of a tin whisker control plan on boards with dual applications. The case study applied the model to two applications: a desktop computer and in a commercial aircraft. This case study was performed to show that if one had accurate data on the PCFC for a system, the cost of performing various activities, and the benefit of performing the same activities, a judgment could be made, with a quantifiable level of certainty, as to the cost-effectiveness of performing the activity in the control plan. In the case study performed for this chapter, performing activities was far more cost effective when the boards were used in a commercial aircraft than when used in a desktop computer, because the equipment had a greater service life requirement and higher financial consequences of failure when used in an aircraft. The boards are projected to fail more often over its service life in an aircraft and the entities responsible for supporting the system incur more cost when there are failures, hence there is more benefit to spending money to reduce the expected number of failures.

References

1. Leidecker H (2006) Tin whiskers: a history of documented electrical system failures. https://nepp.nasa.gov/whisker/reference/tech_papers/2006-Leidecker-Tin-Whisker-Failures.pdf. Accessed 17 Mar 2019
2. Bakhshi R, Kunche S, Pecht M (2014) Intermittent failures in hardware and software. J Electr Packag 136(1):011014–011014
3. Sampson M (2009) Basic info on tin whiskers. Available: https://nepp.nasa.gov/whisker/background/. Accessed 16 Aug 2018
4. Directive 2002/95/EC of the European Parliament (2003) Restriction of hazardous substances in electrical and electronic equipment restriction of hazardous substances in electrical and electronic equipment
5. Han S, Osterman M, Pecht MG (2010) Electrical shorting propensity of tin whiskers. IEEE Trans Electr Packag Manuf 33(3):205–211
6. Ashworth MA, Dunn B (2016) An investigation of tin whisker growth over a 32-year period. Circuit World 42(4):183–196
7. Han S, Osterman M, Pecht M (2011) Likelihood of metal vapor arc by tin whiskers. In: Proceedings of IMAPS advanced technology workshop on high reliability microelectronics for military applications
8. Panashchenko L, Osterman M (2009) Examination of nickel underlayer as a tin whisker mitigator. In: Proceedings of the 59th electronic components and technology conference, pp 1037–1043
9. Mathew S, Osterman M, Shibutani T, Yu Q, Pecht M (2007) Tin whiskers: how to mitigate and manage the risks. In: Proceedings of the international symposium on high density packaging and microsystem integration, pp 1–8
10. Zhang P, Zhang Y, Sun Z (2015) Spontaneous growth of metal whiskers on surfaces of solids: a review. J Mat Sci Technol 31(7):675–698
11. Han S, Osterman M, Meschter S, Pecht M (2012) Evaluation of effectiveness of conformal coatings as tin whisker mitigation. J Electron Mater 41(9):2508–2518

12. Osterman M, Pecht M, Mathew S, Fang T (2006) Tin whisker risk assessment. Circuit World 32(3):25–29
13. Lee B-Z, Lee DN (1998) Spontaneous growth mechanism of tin whiskers. Acta Mater 46(10):3701–3714
14. Fang T, Osterman M, Pecht M (2006) Statistical analysis of tin whisker growth. Microelectron Reliab 46(5):846–849
15. Lillie E, Sandborn P, Humphrey D (2015) Assessing the value of a lead-free solder control plan using cost-based FMEA. Microelectron Reliab 55(6):969–979
16. Barringer P (1997) Life cycle cost & reliability for process equipment. In: Proceedings energy week conference & exhibition, Houston, Texas
17. Sears RW (1991) A model for managing the cost of reliability. In: Proceedings of the annual reliability and maintainability symposium, pp 64–69
18. Jiang ZH, Shu LH, Benhabib B (2000) Reliability analysis of non-constant-size part populations in design for remanufacture. J Mech Des 122(2):172–178
19. Hauge BS, Johnston DC (2001) Reliability centered maintenance and risk assessment. In: Proceedings of the annual reliability and maintainability symposium. 2001 proceedings. International symposium on product quality and integrity (Cat. No.01CH37179), pp 36–40
20. Perera J, Holsomback J (2005) An integrated risk management tool and process. In Proceedings of the IEEE aerospace conference, pp 129–136
21. Sun B, Shu G, Podgurski A, Ray S (2012) CARIAL: cost-aware software reliability improvement with active learning. In Proceedings of the IEEE fifth international conference on software testing, verification and validation, pp 360–369
22. Liu R, Boggs S (2009) Cable life and the cost of risk. IEEE Electr Insul Mag 25(2):13–19
23. Rhee SJ, Ishii K (2003) Using cost based FMEA to enhance reliability and serviceability. Adv Eng Inform 17(3):179–188
24. Kmenta S, Ishii K (2005) Scenario-based failure modes and effects analysis using expected cost. J Mech Des 126(6):1027–1035
25. Taubel J (2011) Use of the multiple severity method to determine mishap costs and life cycle cost savings. In: Proceedings of the 29th international system safety conference, Las Vegas, NV
26. MIL-STD-882C (1993) Department of Defense
27. Sandborn P (2017) Cost analysis of electronic systems, 2nd edn. World Scientific, Singapore
28. Mathew S, Alam M, Pecht M (2012) Identification of failure mechanisms to enhance prognostic outcomes. J Fail Anal Prev 12(1):66–73
29. Das D (1999) Use of thermal analysis information in avionics equipment development. Electron Cooling 5:28–34